CROSS-DISCIPLINARY PERSPECTIVES *on* HOMELAND *and* CIVIL SECURITY

This book is part of the Peter Lang Education list.
Every volume is peer reviewed and meets
the highest quality standards for content and production.

PETER LANG
New York • Bern • Frankfurt • Berlin
Brussels • Vienna • Oxford • Warsaw

CROSS-DISCIPLINARY PERSPECTIVES *on* HOMELAND *and* CIVIL SECURITY

A Research-Based Introduction

EDITED BY **ALEXANDER SIEDSCHLAG**

PETER LANG
New York • Bern • Frankfurt • Berlin
Brussels • Vienna • Oxford • Warsaw

Library of Congress Cataloging-in-Publication Data

Cross-disciplinary perspectives on homeland and civil security:
a research-based introduction / edited by Alexander Siedschlag.
pages cm
Includes bibliographical references and index.
1. Civil defense. 2. National security. 3. Emergency management.
4. Internal security. I. Siedschlag, Alexander.
UA926.C766 363.34'7—dc23 2015014870
ISBN 978-1-4331-2986-5 (paperback)
ISBN 978-1-4539-1579-0 (e-book)

Bibliographic information published by **Die Deutsche Nationalbibliothek**.
Die Deutsche Nationalbibliothek lists this publication in the "Deutsche
Nationalbibliografie"; detailed bibliographic data are available
on the Internet at http://dnb.d-nb.de/.

The paper in this book meets the guidelines for permanence and durability
of the Committee on Production Guidelines for Book Longevity
of the Council of Library Resources.

29 Broadway, 18th floor, New York, NY 10006
www.peterlang.com

Printed in the United States of America

Table of Contents

List of Figures and Tables

Figures

Tables

Acknowledgments

I would like to express my gratitude to a number of people and institutions to whom I owe in the editorship of this book.

My previous experience in shaping the field of civil security research in Europe significantly contributed to the plan for this book: serving the European Commission as an expert in the preparatory action for the Security theme in the 7th Framework Program for research, and later on in that program after it was launched, as well as the Austrian National Security Research Program KIRAS as a founding member of the jury. My years as a Professor for Security Research at Sigmund Freud University Vienna, Austria, and as a coordinator of the EU security research project FOCUS, on "Foresight Security Scenarios," from which several colleagues have contributed to this volume and the concept of the book, were fundamental, too. FOCUS also availed me the privilege of academically collaborating with DHS components and research projects, as well as with FEMA, in their Strategic Foresight Initiative (SFI).

The inspiration and insight I gained on my current appointment as Professor and Chair of Homeland Security Programs at The Pennsylvania State University, at my home college Penn State Harrisburg, and at its School of Public Affairs, has been limitlessly precious. I particularly benefitted from discussion with colleagues and students in the Intercollege Master of Professional Studies Program in Homeland Security (*i*MPS-HLS).

Thank you to the contributing authors; without your insight and contributions this book may not have been possible. I believe our mutual interest, skills, and knowledge lend to a perfect blend and authoring of this volume.

I am grateful for the invaluable support in editing the the entire text that I received from Alison Shuler, B.S. and Lesa Stanford of my Program Office at Penn State Harrisburg. Although they were already swamped with regular duty, they did not hesitate to give of their time to help this project finish on time.

Finally, my bottomless thanks to Andrea, who has been incredibly resilient to and supportive of my promise of never doing a textbook again (at least I switched languages!). She has helped me with literally all of her heart to finish this project among many serious academic responsibilities, and she has been a great editor and peer reviewer as well. Thank you.

List of Abbreviations

ANPR	Automated number-plate recognition
AUVSI	Association for Unmanned Vehicle Systems International
BRP	Bureau of Radiological Protection
BWC	Biological and Toxin Weapons Convention
CBP	U.S. Customs and Border Protection
CBRN	Chemical, biological, radiological, and nuclear
CBRN-E	Chemical, biological, radiological, nuclear, and explosive
CCTV	Closed-circuit television
CEO	Chief executive officer
CIA	Central Intelligence Agency
CIP	Critical infrastructure protection
CSDP	Common Security and Defence Policy (of the European Union)
CVE	Countering violent extremism
CWC	Chemical Weapons Convention
DAESH	see ISIS
DCPA	Defense Civil Preparedness Agency
DEA	Drug Enforcement Administration
DG	Directorate General (of the European Commission)

DHS	U.S. Department of Homeland Security
DNI	Director of National Intelligence
DOD	U.S. Department of Defense
DSCA	Defense Support of Civil Authorities
EC	European Commission (also used for the former European Community, before it completely fused into the European Union with the entry into force of the Lisbon Treaty [2009])
EC3	European Cybercrime Centre
EDA	European Defence Agency
EEODN	European Explosive Ordnance Disposal Network
EGF	European Gendarmerie Force
EOC	Emergency Operations Center
ELINT	Electronic intelligence
ELSI	Ethical, legal, and social issues
EMAC	Emergency Management Assistance Compact
ENLETS	European Network of Law Enforcement Technology Services
EOC	Emergency Operations Center
ERCC	Emergency Response Coordination Centre (EU)
ESF	Emergency Support Function
ESRIF	European Security Research and Innovation Forum
EU	European Union
eu-LISA	European Agency for the Operational Management of Large-Scale IT Systems in the Area of Freedom, Security and Justice
EUROPOL	European Police Office
EUROSUR	European Border Surveillance System
FAA	Federal Aviation Administration
FBI	Federal Bureau of Investigation
FEMA	Federal Emergency Management Agency
FISA	Foreign Intelligence Surveillance Act
FRONTEX	European Agency for the Management of Operational Cooperation at the External Borders of the Member States of the European Union
GDP	Gross domestic product
GEOINT	Geospatial intelligence

GIS	Geographic information systems
GISci/GIScience	Geographic information science
GIT	Geographic information technology
GLONASS	Globalnaya navigatsionnaya sputnikovaya sistema (Global Navigation Satellite System)
GPS	Global Positioning System
GWOT	Global War on Terror
HUMINT	Human intelligence
HVE	Homegrown violent extremism/extremist
IC	Intelligence community
ICE	U.S. Immigration and Customs Enforcement
ICS	Incident Command System
IED	Improvised explosive device
IMINT	Imagery intelligence
INTERPOL	International Criminal Police Organization
INTs	Intelligence collection disciplines
IP	Intellectual property
IPR	Intellectual property rights
ISIL	see ISIS
ISIS	Islamic State of Iraq and Syria or Islamic State of Iraq and al-Sham; the name is also commonly translated as Islamic State of Iraq and the Levant (ISIL); the group's Arabic name is transliterated as *ad-Dawlah al-Islāmīyah fī al-'Irāq wash-Shām* (Da'ish or DAESH)
JIC	Joint Information Center
JTTF	Joint Terrorism Task Forces
MASINT	Measurement and signatures intelligence
MPA	Master of Public Administration
MPS	Master of Professional Studies
NAS	National Airspace System
NATO	North Atlantic Treaty Organization
NGA	National Geospatial-Intelligence Agency
NGO	Non-governmental organization
NIAC	National Infrastructure Advisory Council
NIMS	National Incident Management System
NIPP	National Infrastructure Protection Plan
NPG	National Preparedness Goal
NRC	Nuclear Regulatory Commission

NRF	National Response Framework
NTAS	National Terrorism Advisory System
ODNI	Office of the Director of National Intelligence
OSINT	Open-source intelligence
PEMA	Pennsylvania Emergency Management Agency
PPD	Presidential Policy Directive
PPP	Public-private partnership
PSA	Protective Security Advisor
QHSR	Quadrennial Homeland Security Review
R&D	Research and development
RABIT	Rapid Border Intervention Team
RIP	Resilience implementation process
ROI	Return on investment
RPAS	Remotely piloted aerial system
RTD	Research, technology, and development
SAR	Suspicious activity report
SAT	Structured analytic techniques
SBE	Social, behavioral, and economic sciences
SIEWG	Societal Impact Expert Working Group
SIGINT	Signal(s) intelligence
SIS	Schengen Information System
SLTT	State, local, territorial, and tribal
SWIM	System-wide information management
TCO	Transnational criminal organization
TELINT	Telemetry intelligence
THSGP	Tribal Homeland Security Grant Program
TIGA	Tribally Inclusive Geographic Areas
TMI	Three Mile Island (Nuclear Power Station)
TRL	Technological readiness level
TSA	Transportation Security Administration
UAS	Unmanned aerial system
UAV	Unmanned aerial vehicle
UN	United Nations
UNHCR	United Nations High Commissioner for Refugees
USC	United States Code
WMD	Weapon(s) of mass destruction

Homeland and Civil Security Research Studies for an Evolving Mission Space: Introduction and Overview of Chapters

ALEXANDER SIEDSCHLAG

Studying Homeland Security—How to Know It When You See It

This book offers a research-based introduction to cross-disciplinary perspectives on homeland and civil security, rooted in two basic assumptions: the scholastic assumption that the paradigm of civil security research provides an insightful framework of analysis for homeland security; and the pragmatic assumption that policies, strategies, and programs for homeland security are a subset of a broader effort to ensure civil security, an effort not geographically, culturally, or functionally bound. Homeland security is about risk management in a dynamic all-hazards context that defines its evolving mission space. This does not mean that its objective is to address *all and any* hazards that might emerge. Policies and strategies in the U.S. and elsewhere have pointed out that in order to be effective (and affordable), homeland security needs to be selective, focusing on "the greatest risks" to security,[1] or on those that are responsive to our strategies and technological tools.[2] If the risk-informed approach to prioritizing civil security efforts and resources is followed consistently, we may see practices as well as political and institutional designations change over time because the definition of, and response to, risk is not only evidence based, but also culturally driven.[3] Thus, security is neither implementation of the obvious nor ontological, but an ongoing controversy. A cross-disciplinary perspective is essential for an actionable, balanced view.

This points us to the classic debate about the extent to which security issues are amenable to rational analysis and research, as well as how practically relevant the results of academic inquiry into security can be. For example, writing in 1952, Arnold Wolfers, Professor of Political Science at Yale University, complained that "national security" was a symbol that left too much room for confusion to serve as a guiding principle for political advice or scientific analysis.[4] He suggested that, as a first step in developing an analytical concept of the term, security should be considered, "in an objective sense, [...] the absence of threats to acquired values, [and] in a subjective sense, the absence of fear that such values will be attacked."[5] After the end of the Cold War, security policy continued to be understood as a normative practice, namely as defending values.[6] Today, defending values and the nation's heritage is an important ingredient of homeland security as seen by the U.S. Department of Homeland Security, and reflected in the "Homeland Security Vision" put forward in the *2014 Quadrennial Homeland Security Review*:[7]

> A homeland that is safe, secure, and resilient against terrorism and other hazards, where American interests, aspirations, and way of life can thrive.[8]

This declared vision has considerably broadened the scope of homeland security beyond what was defined during the Bush Presidency in the, still current, *National Strategy for Homeland Security* (2007) as

> a concerted national effort to prevent terrorist attacks within the United States, reduce America's vulnerability to terrorism, and minimize the damage and recover from attacks that do occur.[9]

The state of homeland security as a discipline, and if it is or should be a discipline, remains as debated as what its educational foundations should be.[10] Nevertheless, there is an emerging consensus about what homeland security means as a generic concept that is not owned by any particular nation, although U.S. definitions, policies, and strategies have had defining influence virtually around the world. In the following, the emerging consensus about homeland security as a model for creating security as a public good and deliver it to citizens will be outlined, and it will be explained why and how homeland and civil security are considerably older than the post-9/11 era and the U.S.

The Emerging Homeland Security Consensus

The conceptual connection of "homeland" and "security" clearly pre-dates the terminology framed in the U.S. in response to 9/11 and then adopted or mirrored in other countries, as well as in academia and in technological research and development. Thus, 9/11 was less a "game changer"[11] than a catalyst, not obsolescing but reinforcing the state of the art in what today is known as civil security research. For example, what today is referred to as the homeland security cycle—defined by Presidential Policy Directive 8 (PPD-8) on "National Preparedness" as comprising core capabilities of prevention, protection, mitigation, response, and recovery toward a National Preparedness Goal—[12] draws from what had previously been widely known as the crisis management cycle, an intellectual conclusion from decade-long experience around the world.[13] While homeland security originated as an "Americanized" term, it has evolved into a generic concept.

There is an emerging consensus that *homeland security* is focused on a comprehensive approach, across sectors and tiers of government, driven by a value-based vision that includes a multiphase, multicapability framework with an emphasis on:

- preparedness and the purpose of achieving a resilient nation in an all-hazards perspective, based on risk-management;
- securing borders as well as facilitating lawful flow of people, goods, and information across borders;
- a whole-of-community approach in which homeland security becomes a collective societal responsibility and is regarded as the common creation of a public good;
- delivering security to both the public and private sector, as well as actively including the private sector in the creation of security as a public good; and
- working internationally to realize its broad yet mission-based objectives.[14]

The emerging consensus also includes the view that homeland security is not only about safeguarding borders, infrastructures, and societies, but also about securing values and way of life. Hence, that security needs to be weighed against other values, such as liberty and freedom, but also accountability and deliberation. In this light, critical thinking and continuous improvement are not signs of weakness, but a natural part of the effort and ingredients of success.[15] Lack of critical thinking in homeland security not only can lead to self-serving policies, but to loss of public trust and to mass casualties.[16]

The emerging consensus further includes that homeland security differs from *national security* in that the latter seeks to deter an attack on the country and respond to the aggressor in case an attack occurs.[17] Homeland security, in contrast, cannot deter those who want to threaten, if not eliminate, our way of life because they do not have those rational and strategic purposes that national security concepts have typically seen adversaries guided by. In addition, homeland security is about capabilities to respond and recover internally, a responsibility that had been given to the civil defense sector during the era of Cold War national security. Homeland security integrates those capabilities, making it an endeavor directed at keeping *societies and citizens* safe and secure.[18]

In such an understanding, homeland security is not a subset of national security,[19] but it is of a *transnational* quality. Consequently, the *U.S. National Security Strategy* of 2015 states that homeland security reinforces national security and emphasizes the relevance of "international cooperation" (as opposed to just the U.S. working with its international partners) in homeland security specifically.[20]

There also are indications for an evolving *civil security research* consensus, in favor of contributing to the strengthening of homeland security by providing a cross-disciplinary, "scientifically sound basis for all homeland security efforts," critically assessing what different disciplines have to contribute to meeting the challenge of an all-hazards approach.[21] Contributions have recently focused on technologies for homeland security and associated social issues, such as surveillance,[22] which again underscores the civil security research background of the homeland security mission space. It as well underlines the need for the homeland security field of study to transcend geographic boundaries and analysis of single events and responses, establishing a generic corpus of knowledge and methods. This field of study transcends the largely arbitrary "temporal narrative" that just established 9/11 as a watershed, defining away any experience, knowledge, and methods that existed before.[23] The following sections will detail this argument as well as the concept of civil security research.

From Attica to America: The Homeland and Civil Security Trajectory of History

Homeland security as a whole is older than 9/11, not just its applicable set of research methods and approaches. In a conservative time frame, homeland security is a few hundred years old, and in an epistemological—or knowledge- and evidence-based—time frame, it is millennia old. According to Gus Martin's *Understanding Homeland Security* and in Bruce Maxwell's *Documentary History* of homeland security, it all started in the years of 1797–1801, when the first big U.S. immigration reform was enacted under the Adams administration, known as the Alien and Sedition Acts.[24] These Acts sought to counter a sort of violent extremism, namely domestic unrest and upheaval organized by aliens. The Administration said the Acts strengthened national security during the turbulent times of the French revolutionary wars, but critics maintained that the Acts were actually an effort to suppress public disagreement with the Federalist Party. In his *Documentary History* of homeland security, Maxwell even traces the origins of homeland security back to the year 1607, when in Jamestown (Colony of Virginia), Captain George Kendall, an appointed leader, was put to death for being a spy, inciting a mutiny, and a couple of other charges—with sedition deemed the real cause in retrospective. The jury, at any rate, believed Kendall posed a threat to the colony's homeland, not in terms of territorial integrity, but rather adherence to common values and an emerging way of life among the settlers of diverse backgrounds.[25] This description, so similar to the definition of today's homeland security vision, evidences that the history of the field is, at the very least, older than the United States.

Let us travel further back through history to identify the first roots of homeland security and its relationship to civil security. We arrive during the times of the Delian League, in the 5th century B.C.,[26] where a common-threat environment led different city-states of ancient Greece, the majority of whom were colonies, to explore how to improve their security in what we today would call an all-hazards environment. Comparatively analyzing constitutions of city-states as well as political and social cultures, Aristotle dedicated considerable parts of his *Nicomachean Ethics* and *Politics* to questions of homeland and civil security, including—as we would say today—questions of physical and societal infrastructure protection, and defending the state and its citizens more by upholding a way of life and commonly shared values through a comprehensive and whole-of-community approach, rather than by arms alone.

Aristotle also addressed the question of security in relation to liberty and freedom. It should not surprise us that Aristotle writes in the *Nicomachean Ethics* about height of city walls and in the *Politics* about the moral value of flute playing, and not the other way round. It appears to be for the simple reason that a lack of preparedness as a state of mind requires material compensation (such as strong and high city walls); and that cultivating and defending a city-state, or colony, means to secure a way of life as much as a territory. About two millenniums later, the founding generation of what today is the United States had the same experience because of human nature and sociological basics. In our time, societies around the world care about homeland security because of Aristotelian truths, human nature, and foundational sociological understandings, demonstrating that its essentials have never been owned by one single place or nation, but transmitted throughout centuries and continents.[27] The homeland security enterprise in the U.S. today experiences the same "biopolitical" challenges, including the challenge to connect epistemic insight (i.e., insight based on knowledge, generated by research) with the need to produce and deliver public goods by prudence, taking into account realities and distribution of power.[28]

It also should come as no surprise then that the all-hazards approach and its essential association with homeland security has a history.[29] The same is true of security vs. liberty,[30] as well as preparedness (or "anticipation") and resilience.[31] The same even applies to the very concept of "homeland." Written in 1981, an article in *Millennium: Journal of International Studies* entitled "States and Homelands: The Social and Geopolitical Implications of National Territory"[32] identified cross-national/transversal components of "homeland" that remain a good framework for understanding broader conceptual foundations and research perspectives of homeland security, and how they relate to overarching aspects of civil security research. The proposed conceptual map for the "homeland" included, among others, the components of "folkways" (or way of life) and a value-laden territory (of a nation of commonly acquired values) in the definition of "homeland,"[33] as discussed above. This again points to the civil security research perspective.

Civil Security Research

The concept of civil security relates to a practical effort to making societies safe and secure, as well as to a research paradigm.[34] Civil security emerged after 9/11 in different countries as a concept driving research and policy. There exists a wealth of studies addressing "homeland security" in other countries than the U.S., following approaches informed by the civil security model.[35] Civil security must not be confused with different earlier concepts such as "civil defense"[36] (that dates back before the time of comprehensive emergency management) or defense support of civil authorities[37] (that is one element of comprehensive emergency management and law enforcement, differing from homeland defense as a responsibility of the Department of Defense). Generalizing a definition first elaborated in the European Union (EU), civil security research comprises

> research activities that aim at identifying, preventing, deterring, preparing and protecting against unlawful or intentional malicious acts harming [...] societies; human beings, organisations or structures, material and immaterial goods and infrastructures, including mitigation and operational continuity after such an attack (also applicable after natural/industrial disasters).[38]

Such a conceptual framing of the research dimension of homeland security also is an important step to expanding the scope to further regions in the world, beyond "transatlantic homeland security."[39] It opens up the opportunity to stringently include the Asian-Pacific approach into debates and best practice models. That approach is based on the concept of "human security,"[40] which goes a step further than civil security, representing a multidisciplinary, transnational understanding of security, based on a people-centric view. It was an approach particularly set forth by the United Nations in the 1990s, obviously not with homeland security in mind, but is today followed in particular by countries in the Asia-Pacific region as an approach to realizing national as well as homeland security.[41]

An intellectual precursor of civil security research was Barry Buzan's *People, States, and Fear: The National Security Problem in International Relations.* It was written in 1983 with national security in the Cold War age in mind, but strongly arguing for a "holistic view of security,"[42] which is echoed today in the emphasis on a comprehensive approach and a whole-of-community scope of homeland security, as well as an all-hazards perspective.[43] Applying a civil security research perspective to homeland security helps us increase not only our analytical, but also our practical capability in the field, because it opens up a large repertory of classical knowledge to draw from. This, for example, relates to radicalization of societies and nurturing of extremism,[44] as it does to securing societies from a range of threats (anthropogenic and natural) along the internal-external and the security-defense continua, as well as analyzing societal and cultural responses to those threats. Pitrim A. Sorokin's work on *Man and Society in Calamity: The Effects of War, Revolution, Famine, Pestilence upon Human Mind, Behavior, Social Organization and Cultural Life* of 1942 remains a scholastic vertex in this regard.[45] Applying a wide cross-national focus, analyses like this centered on what today would be called preparedness as a state of mind and a comprehensive approach to resilience.

Important to note, the concept of civil security is not alien to the homeland security debate in the U.S., where it was used to supplant the argument that 9/11 did not make obsolete some relevant state of the art and practical knowledge inherited from the civil defense era, as well as from long-established academic disciplines.[46] The civil security approach taken by this book is strongly supported by the National Research Council study on *Frameworks for Higher Education in Homeland Security.* According to that study, already concluded in 2005, an all-hazards approach and a comprehensive approach covering the full crisis management cycle should be at the heart of all academic programming related to educating future leaders in the field.[47] The study also pointed out that homeland security academic programs should not just echo political concepts and strategic mission spaces, but apply a broad variety of disciplinary perspectives, with "'homeland security' serving more as a target of application of such studies, rather than as a descriptor of the studies themselves."[48] The study additionally highlighted that "[a]cademic inquiry is not constrained by any particular agency's mission, configuration, or programmatic concerns."[49] Within this understanding, the final section of this chapter provides an overview of the various contributions to this volume, along with brief summaries.

Overview of Contributions

This book presents a *research-based* introduction to homeland and civil security, utilizing contributions from across select scholarly as well as professional disciplines that are rooted in current research in the field or research results from the field. These include academic, professional, and policy research; presentation of different example text types in which research results may be disseminated, used, and interpreted to support decision-making; chapters that provide insight in programming for research in homeland and civil security, as well as in assessing the scientific status of the field; and a substantial appendix that offers multiple resources supporting future research. According to the frame of reference outlined so far, the content of this book is derived from select areas of international convergence of policies and programs, approaches, and main security missions in the civil sector.[50] Emphasis was placed on examples from areas particularly driven by or relevant for research that at the same time help improve understanding of overarching aspects and terminology, as well as represent different types of research and styles of writing.

The book consists of the following six sections:

- Aspects from the Mission Space (Part I);
- Aspects from Disciplines (Part II);
- Transversal Aspects (Part III);
- Aspects of Programming for Homeland and Civil Security (Part IV);
- Conclusion and Outlook (Part V); and
- Practical Appendix with Glossary, Bibliography, Learning and Research Guide, and Index (Acronyms are explained in the front matter).

This section provides an overview of the chapters.

Part I: Aspects from the Mission Space

Chapter 1: Examining the Strategic Hybrid Threat: Technology, Terrorism, Transnational Criminal Organizations, and Old Enemies after 2015

In Chapter 1, "Examining the Strategic Hybrid Threat: Technology, Terrorism, Transnational Criminal Organizations, and Old Enemies after 2015," Robert McCreight presents a long-served U.S. Department of State expert's view on global security issues and how they relate to, or translate into, homeland security threats. The chapter is also an example of the executive writing style typical of that part of the professional community, and a practical exemplification of what exploration of the external-internal security continuum means from the national security professional's point of view. Based on the reality of limited national and international capabilities, the chapter argues that after a decade of focus on the Global War on Terror (GWOT), we have to target strategies and resources abroad and at home to directly addressing the multidimensional threat landscape, based on continuous, honest, and futuristic assessment.

Chapter 2: Border Management: International Experiences

Border management is a generic civil security capability. It plays a prominent role in virtually any democratic country's effort to create security as a public good and deliver it to its citizens. In Chapter 2, "Border Management: International Experiences," Nadav Morag provides a comparative analysis of main factors in border management and how they are approached by the European Union (and non-EU Schengen countries), Australia, and Israel. This contribution is an example of a public administration-based approach to homeland and civil security, focusing on how border management policies are translated into strategies and practices. The chapter also introduces border management as a comprehensive concept that includes general border control, as well as specific measures for coping with immigrants, asylum claimants, and refugees; for enhancing security and coping with cross-border threats; and for coordinating activities between and among domestic and foreign agencies.

Chapter 3: The Three Mile Island Nuclear Disaster from an Emergency Management Perspective

Emergency management has been a part of the homeland security mission space ever since the Department of Homeland Security was created in 2002. The Federal Emergency Management Agency (FEMA) then became a component of the U.S. Department of Homeland Security. FEMA was activated very shortly after the accident in the nuclear power station of Three Mile Island (TMI) in March 1979, within sight of where, on The Pennsylvania State University's Penn State Harrisburg campus, this book has been edited and this introductory chapter written.[51] In Chapter 3, "The Three Mile Island Nuclear Disaster from an Emergency Management Perspective," Kevin J. Molloy shares his first-hand experience with TMI emergency management in 1979, drawing a series of insightful conclusions for today's era of homeland security. The chapter represents, in an anecdotal case-study type of writing, an accomplished practitioner's analysis with structured conclusions pointing out lessons learned as well as challenges persisting in our era of homeland security.

Chapter 4: Beyond the Storms: Implementing Smart Resilience in Turbulent Times

In Chapter 4, "Beyond the Storms: Implementing Smart Resilience in Turbulent Times," Dane S. Egli presents a scholarly analysis of what resilience means conceptually and in practice. While focusing his practical analysis on the United States and its critical infrastructure protection needs, Egli provides an international state-of-the-art perspective on resilience as a foundational concept for civil security, arguing that resilience is "disaster agnostic." It requires programs and strategies not bound to specific points in time, on the map, or in a mindset. The chapter also addresses the role of global supply chains and offers a useful model to overcome the limitations of a perimeter approach to critical infrastructure protection.

Part II: Aspects from Disciplines

Chapter 5: The Role of Intelligence in Homeland Security

Chapter 5, "The Role of Intelligence in Homeland Security," by William J. Ryan explains from the homeland security professional's point of view why intelligence is an integral ingredient in the effort to comprehensively realize the mission of homeland security—and how high-quality intelligence influences all mission areas of homeland security (prevention, protection, mitigation, response, and recovery) across the different tiers of government. Ryan argues that intelligence for homeland security is a generic concept and reaches beyond the U.S. because, by appropriate collection and use of intelligence, countries around the world are able to develop and implement better homeland security. The chapter uses illustrative scenarios and emphasizes the necessity of partnerships in collecting, disseminating, and using intelligence.

Chapter 6: Geospatial Intelligence and the Geospatial Revolution

Based on the general framework for homeland security intelligence described in Chapter 5, Chapter 6, "Geospatial Intelligence and the Geospatial Revolution," by Todd S. Bacastow and Gregory A. Thomas introduces Geospatial Intelligence (GEOINT) from a scholarly perspective as "the art and science of place and time on Earth's surface. Its subject matter is the human phenomena that make up the world's places." This discipline has evolved from military and defense realms to homeland security, including emergency management and law enforcement, as well as to other fields, such as business and international humanitarian efforts. The chapter illustrates how what the GEOINT professional does has become directly relevant to an increasingly internationalizing homeland security enterprise.

Chapter 7: Use of "Drones" in Homeland Security: A Comparative Perspective on a Security Technology and Its Legal, Political, and Social Aspects

Chapter 6 on geospatial intelligence already included, among other things, reference to use of "unmanned aerial vehicles (UAVs)." This topic is expanded on in Chapter 7, "Use of 'Drones' in Homeland Security: A Comparative Perspective on a Security Technology and Its Legal, Political, and Social Aspects." Written by Andrea Jerković and Alexander Siedschlag as an example of a research-report type of text, crafted in executive style, it offers a pragmatic, comprehensive overview of an important and evolving field of technology in homeland and civil security, and its implications. Concentrating on a U.S.-EU comparison, this includes comparative legal and political framework analysis, also assessing public perception and acceptance aspects, as well as ethics and risk-management issues.

Chapter 8: Risk Management within Tribally Inclusive Geographic Areas: A Primer for All Risk Managers

Chapter 8, "Risk Management within Tribally Inclusive Geographic Areas: A Primer for All Risk Managers," by Lloyd Mitchell discusses defining and describing tribal communities; risk management in Tribally Inclusive Geographic Areas (TIGA); and reducing economic impacts of intentional, variably intentional, or unintentional terror or disaster events by working together within TIGA. The chapter emphasizes the relevance of a perspective on the tribal level of homeland security governance (the final "T" in the often-cited SLTT tier, standing for state, local, territorial, and tribal levels, in addition to the federal level) and risk management, based on appropriate practical use of terminology in addition to use of resources specifically made available to tribes.

Part III: Transversal Aspects

Chapter 9: The Leadership LEAP: A New Approach for Homeland Security

Much is said about the need for leadership in homeland security, and the need for related academic programs to support development of leaders. Leadership also is a topic of broad relevance in civil security writ large. While crises are self-organizing, complex, and emerging, those who tackle them do not simply emerge, but need to be developed into leaders before the crisis. In Chapter 9, "The Leadership LEAP: A New Approach for Homeland Security," R. Jeffery Maxfield, Rodger Broomé, and John R. Fisher present a supporting model, drawing from a large set of literature and knowledge from leadership research. While management is an important function in any operation, controlling activities of systems and things, leadership is the influence of relationships among people. Most important leader traits include the need to be: *legitimate* and have authentic power; able to make *ethical* decisions when confronted with choices and/or dilemmas, and under stress; *affective,* with the ability to create as well as have trust, and create vision with others; and *persistent,* showing determination in achieving goals and objectives. Legitimate-ethical-affective-persistent then gives the model the acronym "LEAP."

Chapter 10: Ethical, Legal, and Social Issues (ELSI) in Homeland and Civil Security Research and the European Union Approach

While leadership, as addressed in Chapter 9, is an important transversal aspect when it comes to practical competencies, addressing of ethical, legal, and social issues (also known as ELSI) is a leadership competence as well as a quality component in research programs and projects. Chapter 10, "Ethical, Legal, and Social Issues (ELSI) in Homeland and Civil Security Research and the European Union Approach," by Katerina Hadjimatheou, Tom Sorell, and John Guelke explains the ELSI challenge in general and then illustrates how the European Union addresses it, citing select examples from security research projects. Those include the need to balance liberty and security in the face of an all-hazards threat environment; to address specific ethical, legal, and social issues arising with the use and export of security technologies; and to use appropriate technical and regulatory means in order to solve those issues.

Part IV: Aspects of Programming for Homeland and Civil Security

Chapter 11: Homeland Security within a School of Public Affairs

Chapter 11, "Homeland Security within a School of Public Affairs," by Steven A. Peterson is a genealogical essay about The Pennsylvania State University's Intercollege Master of Professional Studies in Homeland Security program (*i*MPS-HLS).[52] Having its academic home at the University's School of Public Affairs, it includes five other disciplines, preparing students for careers across the homeland security enterprise, centered on a comprehensive approach. The political science and public administration perspective on homeland security is particularly valuable to address homeland security governance aspects across tiers (federal, state, local, territorial, and tribal); across public and private entities; in response to an all-hazards threat environment; and with a focus on risk management, since homeland security cannot create an ontology of total protection.

Chapter 12: Trends on Security Research in Europe

The events of September 11, 2001 not only brought about political, legal, and institutional changes in the U.S., but also in other parts of the world. Founded in 1993 as an integration of Western European institutions created in the 1950s, the European Union (EU) responded to 9/11 with an effort to expand on its Member States' capabilities to adapt to the new threat environment by technological innovation and a common response to radicalization. Following the Treaty of Lisbon that entered into force in 2009, the EU has legal personality, and it also has an *Internal Security Strategy*, adopted in 2010, advocating a "European Security Model."[53] An increasing effort has been placed on enhancing Member State capabilities and creating European-level policies and capabilities to deliver security to the citizens, with all citizens of the Member States also having European Union citizenship. This effort is supported by EU programming for civil security research, rooted in an all-hazards approach. In Chapter 12, "Trends on Security Research in Europe," Michel F. Bosco, who leads security research at the European Commission, describes how the European security research program developed in response to a general vision to enhance resilience, to respond to natural and anthropogenic disasters, to bridge the hiatus between civilian and military research and technology development, as well as to balance liberty vs. security. The chapter also introduces relevant EU institutions and agencies in charge of homeland security, known in the EU as "home affairs," and how the security research program is responding to and supporting their security policies and strategies. This provides general insight into the bandwidth of policy areas and agencies involved in civil security.

Part V: Conclusion and Outlook

Chapter 13: The Scientific Status of New Security Studies: A Critical Search for Epistemic Identity of Homeland and Civil Security Research

The concluding Chapter 13, "The Scientific Status of New Security Studies: A Critical Search for Epistemic Identity of Homeland and Civil Security Research," by Rachel Suissa offers a concluding critical assessment of the field, discussing the quality of scientific inquiry into homeland and civil security from an *epistemic* point of view—the character of the knowledge it produces. The chapter puts the cross-disciplinary perspectives on homeland and civil security offered in this book into the overarching context of *New Security Studies,*[54] a field defined by four pillars: new security concepts (e.g., hybrid threats and resilience); new security subjects (e.g., security as ethics); new security objects (e.g., borders and industrial accidents); and new security practices (e.g., practices of technology use or professional security practices such as leadership).

As the homeland security mission space, the civil security model, and the scholarly and political debates continue to evolve, it will be essential for the field to maintain a comprehensive perspective on its subject matter, as well as on its own shape as a research enterprise fed by different disciplines, including academic disciplines as well as professional disciplines, or communities of practice.

Notes

1. Presidential Policy Directive 8 (PPD-8), National Preparedness, March 30, 2011, http://www.dhs.gov/presidential-policy-directive-8-national-preparedness; U.S. Department of Homeland Security, *National Preparedness Goal,* 1st ed., September 2011, http://www.fema.gov/media-library-data/20130726-1828-25045-9470/national_preparedness_goal_2011.pdf
2. European Security Research and Innovation Forum (ESRIF), *ESRIF Final Report* (Brussels, 2009), http://ec.europa.eu/enterprise/policies/security/files/esrif_final_report_en.pdf
3. Mary Douglas and Aaron Wildavsky, *Risk and Culture: An Essay on the Selection of Technological and Environmental Dangers* (Berkeley and Los Angeles, CA: University of California Press, 1983); Robert Wuthnow, *Be Very Afraid: The Cultural Response*

to Terror, Pandemics, Environmental Devastation, Nuclear Annihilation, and Other Threats (Oxford: Oxford University Press, 2010).

4. Arnold Wolfers, "'National Security' as an Ambiguous Symbol," *Political Science Quarterly* 67 (1952): 483.
5. Ibid., 485.
6. Barry Buzan, *People, States, and Fear: An Agenda for International Security Studies in the Post-Cold War Era* (Boulder, CO: Rienner, 1991).
7. U.S. Department of Homeland Security, *The 2014 Quadrennial Homeland Security Review* (Washington, D.C., June 2014), 14, http://www.dhs.gov/sites/default/files/publications/2014-qhsr-final-508.pdf
8. Ibid.
9. The President of the United States, Homeland Security Council, *National Strategy for Homeland Security* (Washington, D.C.: The White House, October 2007), 3, http://www.dhs.gov/xlibrary/assets/nat_strat_homelandsecurity_2007.pdf
10. Michael D. Falkow, "Does Homeland Security Constitute an Emerging Academic Discipline?" (Master's thesis, Naval Postgraduate School, 2013), https://www.hsdl.org/?view&did=736325; William V. Pelfrey and William D. Kelley, "Homeland Security Education: A Way Forward," *Homeland Security Affairs* 9 (2013), https://www.hsaj.org/articles/235; as well as the discussion in the *Journal of Homeland Security Education,* http://www.journalhse.org
11. Jonathan Paquin and Patrick James, eds., *Game Changer: The Impact of 9/11 on North American Security* (Vancouver and Toronto: UBC Press, 2014).
12. Presidential Policy Directive 8 (PPD-8); U.S. Department of Homeland Security, *National Preparedness Goal.*
13. "Crisis Management Cycle," *Urban Securipedia,* last published August 15, 2013, http://securipedia.eu/mediawiki/index.php/Crisis_management_cycle
14. This consensus is summarized in Mike Bourne, *Understanding Security* (Houndmills, Basingstoke: Palgrave Macmillan, 2014), chapter 4: "The Sovereign State and Internal Security" (71–92). It is mirrored by recent handbooks and other publications from the field, as compiled in the "Homeland Security Learning and Research Guide" at the end of this book (281–296).
15. Cf. Kathiann M. Kowalski, *A Pro/Con Look at Homeland Security: Safety vs. Liberty after 9/11* (Berkeley Heights, NJ: Enslow, 2008); see also Frank P. Harvey, *The Homeland Security Dilemma: Fear, Failure and the Future of American Insecurity* (New York: Routledge, 2008).
16. James Ramsay and Linda Kiltz, "Introduction," in *Critical Issues in Homeland Security: A Casebook,* ed. James D. Ramsay and Linda Kiltz (Boulder, CO: Westview, 2014), 2.
17. Cf. Andrew Dorman and Joyce Kaufman, eds., *Providing for National Security: A Comparative Analysis* (Stanford, CA: Stanford University Press, 2014).

18. Jane A. Bullock, George D. Haddow, and Damon P. Coppola, *Introduction to Homeland Security: Principles of All-Hazards Risk Management,* 4th ed. (New York: Elsevier, 2013); Carme Ferro, David Henry, and Thomas MacLellan, *A Governor's Guide to Homeland Security* (Washington, D.C.: National Governors Association [NGA], NGA Center for Best Practices, Homeland Security & Public Safety Division, November 2010), http://www.nga.org/files/live/sites/NGA/files/pdf/1011GOV GUIDEHS.PDF
19. As suggested by Jerome H. Kahan, "What's in a Name? The Meaning of Homeland Security," *Journal of Homeland Security Education* 2 (2013), http://www.journalhse.org/sft710/kahanjhsearticlefinal2.pdf
20. The President of the United States, *National Security Strategy* (Washington, D.C.: The White House, February 2015), 14–15, https://www.whitehouse.gov/sites/default/files/docs/2015_national_security_strategy.pdf
21. Sandra F. Amass et al., eds., *The Science of Homeland Security* (West Lafayette, IN: Purdue University Press, 2006), 7; see also John G. Voeller, ed., *Wiley Handbook of Science and Technology for Homeland Security* (Hoboken, NJ: Wiley, 2010).
22. Francesco Flammini, Roberto Setola, and Giorgio Franceschetti, eds., *Effective Surveillance for Homeland Security: Balancing Technology and Social Issues* (Boca Raton, FL: CRC Press, 2013).
23. Columba Peoples and Nick Vaughan-Williams, *Critical Security Studies: An Introduction,* 2nd ed. (London and New York: Routledge, 2015), "Introduction: Mapping Critical Security Studies and Travelling without Maps" (1–12), 8; see also John W. Collins Jr. et al., "Creating an Executive Doctorate in Civil Security in the United States," *Journal of Homeland Security and Emergency Management* 9, no. 2 (2012).
24. Gus Martin, *Understanding Homeland Security* (Los Angeles, CA: Sage, 2015), 352; Bruce Maxwell, *Homeland Security: A Documentary History* (Washington, D.C.: CQ Press, 2004), 1–6.
25. Maxwell, *Homeland Security*, xxv.
26. "Delian League," *Wikipedia,* last published March 24, 2015, http://en.wikipedia.org/wiki/Delian_League
27. Cf. James Beckman, *Comparative Legal Approaches to Homeland Security and Anti-Terrorism* (Aldershot, Hampshire: Ashgate, 2007), ix.
28. Cf. Christos Boukalas, *Homeland Security, Its Law and Its State: A Design of Power for the 21st Century* (New York: Routledge, 2014).
29. See Allen H. Barton, *Communities in Disaster: A Sociological Analysis of Collective Stress Situations* (Garden City, NY: Doubleday, 1969); Russell R. Dynes and Kathleen J. Tierney, eds., *Disasters, Collective Behavior, and Social Organization* (Newark: University of Delaware Press, 1994).
30. Harold D. Lasswell, *National Security and Individual Freedom* (New York: McGraw-Hill, 1950).

31. See Aaron Wildavsky, *Searching for Safety* (New Brunswick, NJ: Transaction Books, 1988), 168–188 and 220–221; see also Leon H. Weaver, *The Civil Defense Debate: Differing Perceptions of a Persistent Issue in National Security Policy* (East Lansing, MI: Michigan State University, Social Science Research Bureau, 1967).
32. Anthony D. Smith, "States and Homelands: The Social and Geopolitical Implications of National Territory," *Millennium: Journal of International Studies* 10, no. 3 (1981).
33. Ibid.
34. Security research as a *"paradigm"* includes a common understanding of the subject matter of civil security; patterns of consensus how to address security in research, including a catalogue of best practices from research; and communities of research and practice that carry the approach further. This understanding of the term "paradigm" follows the classical definition of Thomas Kuhn, *The Structure of Scientific Revolutions* (Chicago, IL: University of Chicago Press, 1962); see also Margret Masterman, "The Nature of a Paradigm," in *Criticism and the Growth of Knowledge*, eds. Imre Lakatos and Alan Musgrave (Cambridge: Cambridge University Press, 1965). Seminal contributions to the civil security research paradigm include: European Security Research and Innovation Forum (ESRIF), *ESRIF Final Report*; Robert J. Fischer, Edward P. Halibozek, and David C. Walters, *Introduction to Security*, 9th ed. (Waltham, MA: Butterworth-Heinemann, 2013); Martin Gill, ed., *The Handbook of Security*, 2nd ed. (Basingstoke: Palgrave Macmillan, 2014); Clifton L. Smith and David J. Brooks, *Security Science: The Theory and Practice of Security* (New York: Elsevier, 2013); Klaus Thoma, ed., *European Perspectives on Security Research* (Munich: acatech—Deutsche Akademie der Technikwissenschaften, 2011), http://www.acatech.de/fileadmin/user_upload/Baumstruktur_nach_Website/Acatech/root/de/Publikationen/Englisch/acatech_diskutiert_European-Research_WEB.pdf; Voeller, *Wiley Handbook of Science and Technology for Homeland Security;* Lucia Zedner, *Security* (London: Routledge, 2009).
35. Major contributions include: Martin J. Alperen, *Foundations of Homeland Security: Law and Policy* (Hoboken, NJ: Wiley, 2011); Beckman, *Comparative Legal Approaches*; Anja Dalgaard-Nielsen and Daniel S. Hamilton, eds., *Transatlantic Homeland Security: Protecting Society in the Age of Catastrophic Terrorism* (London et al: Routledge, 2006); Peter Faber and Valeri Ratchev, eds., *Homeland/Societal Vulnerability and Security*. Special issue, *Information & Security* 17 (2005), http://procon.bg/17-homeland-societal-vulnerability-and-security; Giorgio Franceschetti and Marina Grossi, eds., *Homeland Security: Threats, Countermeasures, and Privacy Issues* (Norwood, MA: Artech House, 2011); Christian Kaunert, Sarah Leonard, and Patryk Pawlak, eds., *European Homeland Security: A European Strategy in the Making?* (New York: Routledge, 2012); Jeffrey A. Larsen and Tasha L. Pravecek, *Comparative U.S.-Israeli Homeland Security*, Counterproliferation Paper, no. 34 (Maxwell Air Force Base, AL: Air University, United States Air Force Counterproliferation Center, 2006), http://www.au.af.mil/au/awc/awcgate/cpc-pubs/larsen3.pdf; David A. McEntire, ed., *Comparative Emergency Management: Understanding Disaster Policies, Organizations, and*

Initiatives from Around the World (Emmitsburg, MD: FEMA Emergency Management Institute, 2013), https://training.fema.gov/hiedu/aemrc/booksdownload/compemmgmtbookproject; Nadav Morag, *Comparative Homeland Security: Global Lessons* (Hoboken, NJ: Wiley, 2011); Rick "Ozzie" Nelson, project director, *U.S.-India Homeland Security Cooperation: Building a Lasting Partnership via Transportation Sector Security.* Center for Strategic and International Studies: A Report for the CSIS Homeland Security and Counterterrorism Program (Lanham, MD: Rowman & Littlefield, 2013); Gert Vermeulen and Wendy De Bondt, *Justice, Home Affairs and Security: European and International Institutional and Policy Development* (Antwerpen: Maklu, 2015).

36. "Civil Defense," *Wikipedia,* last published March 14, 2015, http://en.wikipedia.org/wiki/Civil_defense
37. "Defense Support of Civil Authorities," *Wikipedia,* last published June 20, 2014, http://en.wikipedia.org/wiki/Defense_Support_of_Civil_authorities; see also the *Homeland Defense and Civil Support Journal* published by the U.S. Army War College, http://www.csl.army.mil/homelanddefensejournal/JournalHome.aspx
38. European Security Research Advisory Board (ESRAB), *Meeting the Challenge: The European Security Research Agenda. Report from the European Security Research Advisory Board* (Luxembourg: Office for Official Publications of the European Communities, 2006), http://ec.europa.eu/enterprise/policies/security/files/esrab_report_en.pdf
39. Studies in transatlantic homeland security for example include: Kristin Archick, coord., "European Approaches to Homeland Security and Counterterrorism," Congressional Research Services Report for Congress, RL33573, July 24, 2006, http://www.fas.org/sgp/crs/homesec/RL33573.pdf; Esther Brimmer, ed., *Transforming Homeland Security: U.S. and European Approaches* (Washington, D.C.: Center for Transatlantic Relations, 2006); John L. Clarke, ed., *Armies in Homeland Security: American and European Perspectives* (Washington, D.C.: National Defense University Press, 2006); Dalgaard-Nielsen and Hamilton, *Transatlantic Homeland Security.* For major contributions to the study of homeland security beyond the transatlantic region, see note 35.
40. "Human Security," *Wikipedia,* last published March 28, 2015, http://en.wikipedia.org/wiki/Human_security; "Human Security Report Project," last accessed March 29, 2015, http://www.hsrgroup.org
41. See William T. Tow, David Walton, and Rikki Kersten, eds., *New Approaches to Human Security in the Asia-Pacific: China, Japan and Australia* (Farnham, Surrey: Ashgate, 2013).
42. Barry Buzan, *People, States, and Fear: The National Security Problem in International Relations* (Chapel Hill, NC: The University of North Carolina Press, 1983), "Conclusion: A Holistic View of Security," 245–258.
43. See U.S. Department of Homeland Security, *The 2014 Quadrennial Homeland Security Review.*

44. Cf. William Kornhauser, *The Politics of Mass Society* (Glencoe, IL: Free Press, 1959).
45. Pitrim A. Sorokin, *Man and Society in Calamity: The Effects of War, Revolution, Famine, Pestilence upon Human Mind, Behavior, Social Organization and Cultural Life* (New York: Dutton, 1942); see also George W. Baker and Dwight W. Chapman, eds., *Man and Society in Disaster* (New York: Basic Books, 1962).
46. Amanda J. Dory, *Civil Security: Americans and the Challenge of Homeland Security* (Washington, D.C.: Center for Strategic and International Studies, 2003); Collins et al., "Creating an Executive Doctorate in Civil Security in the United States."
47. National Research Council of the National Academies, Committee on Educational Paradigms for Homeland Security, *Frameworks for Higher Education in Homeland Security* (Washington, D.C., 2005), 3–4.
48. Ibid., 4.
49. Ibid.
50. This convergence has been discussed by contributions such as Brimmer, *Transforming Homeland Security*; Morag, *Comparative Homeland Security*; or Dalgaard-Nielsen and Hamilton, *Transatlantic Homeland Security*.
51. See also "TMI@35: The Three Mile Island Nuclear Crisis in Perspective. Analyses, Stories, Policies. Documentation of a Conference Held at Penn State Harrisburg, March 27–28, 2014," Penn State Homeland Security Programs, last accessed March 29, 2015, http://sites.psu.edu/tmi35
52. See "Master of Professional Studies in Homeland Security," *Penn State World Campus*, last accessed March 29, 2015, http://www.worldcampus.psu.edu/degrees-and-certificates/homeland-security/overview
53. Council of the European Union, "Draft Internal Security Strategy for the European Union: 'Towards a European Security Model,'" Brussels, March 8, 2010, http://register.consilium.europa.eu/pdf/en/10/st07/st07120.en10.pdf. The "European Security Model" refers to an integrated security architecture containing common tools for collaboration between law enforcement authorities and judicial bodies within the EU Member States. It also includes addressing the causes of insecurity and not just its effects, with priorities on prevention across sectors (political, economic, social, etc.), resting on the assumption that internal security needs to be seen as a comprehensive concept that cuts across various sectors in seeking to address major threats.
54. Cf. J. Peter Burgess, ed., *The Routledge Handbook of New Security Studies* (Milton Park: Routledge, 2010).

Part I

Aspects from the Mission Space

1

Examining the Strategic Hybrid Threat: Technology, Terrorism, Transnational Criminal Organizations, and Old Enemies after 2015

ROBERT MCCREIGHT

Introduction

In our turbulent times, it makes sense to question whether the next decade will be as fraught with tragedy, risk, warfare, terrorism, and insurrection as the last decade seemed to be. Dead reckoning informs us that nations equipped to thwart and destroy future threats are in a much stronger position than states who cannot mount a retaliatory package strong enough to deter or smite future attackers. We know intuitively that aggressive acquisition of power, territory, and primeval violence draws its resolve from mythic determination and incrementally rewarded excursions into realms of asymmetric and destabilizing warfare. Weaker states, those already hobbled by corruption, infrastructural failure, and uncertain leadership can easily fall prey. In some cases, their vulnerability affects our own vulnerability, both strategic and homeland-related, and their paucity of robust military strike options similarly impairs our own. It behooves the United States to assess weakened, crumbling, and utterly failed states in the next decade, for it is from some of these forlorn places that new threats will arise.

The strategic dilemma grows more intense as we ponder the choices before us. Greater scrutiny is invited to determine which of these hapless states merit our intervention and assistance and which do not. It is unlikely that we will see fewer failing states in the next decade and it appears clear we may discover many others reluctant or incapable of defending their land and people from a mix of internal and external threats. The globe has seen sea changes in many areas and for every instance of progress and stability, there is a fragile state poised mercilessly on the brink of failure. The usual mixture of conflict, famine, disease, upheaval, and disaster seem affixed to the landscape of nations clinging tenaciously to some form of political and economic legitimacy.

Of course, the impending global threat remains steadfastly rooted in many older attack concepts such as the infamous "Bojinka" plot from years ago. Hatched by Ramzi Youseff in the mid-1990s, it featured a plan to explode several airliners as they transited the expansive Pacific Ocean. A proven concept remains attractive. In fact, claims of a modified Bojinka plot involving the simultaneous explosion of several European jetliners inside the holiday season at the end of 2014 comes to mind.[1] This alleged "Christmas Spectacular," as referred to in the media, is a throwback to earlier terrorist undertakings, but it is a sobering reminder that mixing older methodologies with newer approaches never goes out of style.

To determine whether the next decade is more chaotic and laden with mayhem than the one we left behind, we need to ask ourselves whether the core elements of a global threat have changed. My argument is that indeed they have. Not only is there reliance on older, tried-and-true attack options, but newer ideas continue to emerge. Therefore, one prominent exhibit worth examining is the evolution of the strategic hybrid threat—at least as I define it.

Examining the Strategic Hybrid Threat for the Next Decade

The strategic hybrid threat outlined for the period 2015–2025 is a mixture of old and new issues which taken separately pose their own strategic challenges and dilemmas.[2] However, when taken together they represent a wide spectrum of threat issues that may exceed the ability of the United States to overcome and reckon with by itself, raising the serious concern about always seeking some form of coalition response, reaction, or retaliation. This is especially true given the steady decline of wider international support from treaty allies, likeminded countries, or the United Nations community over the next decade given a combination of limited resources, domestic political pressures, global energy trade exigencies, and the failure of leaders and policymakers to articulate and define a valid endgame for any future campaigns.

Added to this dilemma is the overall drawdown of forces from Iraq and Afghanistan against a minor surge to deal with ISIS threats in the last week of

2014 and the diminished Department of Defense budget itself. For example, at the end of 2013 with further cuts looming, Army Chief of Staff General Ray Odierno mentioned that the armed forces might even have to submit an "unfunded requirements" wish list to Congress for the first time in years. He went on to state, "I worry about the unknown contingency. We'll continue to train for our Afghan mission and some other missions we have, but for unknown contingencies, our risk goes way up." Moreover, he stressed, "we've got to be prepared to operate across a broader spectrum of conflict, and that's what makes this even more challenging."[3]

The old issues carried over from the period 1954–2014 embrace a collection of viable threats that have plagued the United States over six decades and show no sign of withering away or diminishing in importance. The next decade in particular raises a number of serious issues where alignment between the United States and its traditional allies may be tested. Aside from the perennial question of securing appropriate resources, there are major unanswered questions about advanced technologies such as "drones" and acoustic wave systems, which have both offensive and defensive characteristics that can be exploited by friends and foes alike.

In perspective, we must begin to recognize the array of genuine threats that our national security policy must address. Calibrated as it must be for the spectrum of well-known, anticipated, and likely attacks on our people, territory, and sovereignty, it must also leave room for the unexpected. Naturally, this would entail a diabolical blend of well-known, routinely suspected and newly emerging threats that cannot easily be dismissed or readily ignored. A number of valid candidates come to mind.

This would include classical super power rivals like Russia and China, long-standing states of concern like North Korea and Iran, as well as those countries whose strategic viewpoints are frequently at odds with the United States—like Brazil, Peru, and Indonesia. It also must include states where the regime is in jeopardy or adrift such as Syria, Sudan, and Somalia as well as problematic states that pose unique and compelling security dilemmas, such as Egypt, Saudi Arabia, Venezuela, and Pakistan. The dictum that advises "be ready for anything" may be entirely realistic, but it stretches the resources and capacity of our global and national defense beyond simple readiness and geostrategic threat discernment. These nations sustain leaders and economies that openly trade with and seek mutually supportive arrangements with benign states, pariah states, and some of our old enemies. They do so fully aware that this sometimes pokes a geostrategic finger in Uncle Sam's eye and it temporarily enhances their own stature as a state that actually stood up to America's alleged unquenchable appetite for influence and power. They distance themselves from Washington, knowing full well that this stance affords them leverage and prestige they would never otherwise derive. Worse, maneuvering for more global status and opting to leverage their regional and territorial interests, they compound the issue of ensuring that marginally stable states can face internal and external threats to their homelands' security with aplomb.

For example, seeking to sustain a stable and secure tri-ethnic Iraq after our military drawdown is an outcome that should be pleasing to most of Iraq's neighbors. However, this is not easily obtained. It is conceivable that alternative scenarios may emerge where it seems more likely that Iraq might one day fragment and dissolve along ethnic and territorial lines or collapse into renewed sectarian strife. Is this, in fact, more geopolitically pleasing to Iran, Turkey, and several of our reliable Middle Eastern associates than an Iraq that remains a viable and peaceful sovereign state? It raises the question of what the "ideal" status of Iraq should be for the next decade. Is it one that somehow balances U.S. interests with those of Iraq's territorial neighbors and trading partners? Or is it instead simply a sovereign but ungovernable territory?

As if that collection of challenges were not hard enough, we now must deal with another layer of complexity. New issues, relatively speaking, include the twin threats of terrorist groups and the growing political and economic interests of transnational criminal organizations (TCOs), which have steadily enhanced their leverage and influence through a blend of violence, social media, financial dealings, and the covert acquisition of improved armaments. Their opportunities for cooperation, collaboration, and explicit commonwealth may seem far-fetched, but we overlook this risk at our peril.

Along with these two threats, which display no evidence of diminished influence, is the growth of dual-use technologies like cyber, neuroscience, nanoscience, genomics, robotics, lasers, acoustic wave, and other cutting-edge scientific endeavors, which respect no national boundaries, and which contain a largely unassessed potential for unleashing new forms of future weapons. The price we pay for being seduced by advanced technologies is that we often forget they can sting as much as they reward. Worse, we must accept that our own views of actual, incipient and implied threats is one that may not be shared by our friends and allies to the extent of danger and fascination we hold true. When there is cleavage between our nation and our allies, it is fair to ask whether this is a natural byproduct of interstate relations, or whether differences have been engineered subtly by those who stand to gain the most.

Homeland Security Implications of the Strategic Hybrid Threat

Since the 9/11 attacks where the bulk of our efforts in the decade afterwards were focused primarily on the Global War on Terror (GWOT), we have come to the realization that dividing our overall strategic efforts toward a mix of conventional threats, insurrections, and state instability and incipient terrorism may not be enough. Instead, we have to reorient our perspective toward a multidimensional threat landscape that has profound implications for homeland security.

As we examine the serial and parallel threats to our infrastructures, our commercial aviation, our borders, our maritime interests, and our cyber systems, which have grown daily in sophistication since 2010, we must reckon with the extent to which accelerated technologies, if weaponized, would alter the security landscape and greatly complicate the homeland security mission. Greater technical research and development in areas such as synthetic biology, robotics, artificial intelligence, nanoscience, and neuroscience, for example, over the next crucial decade after 2015 contain significant dual-use risks that potentially could be exploited by traditional enemies, states of concern, terrorists, TCOs, or a combination of these groups. What they share in common is a desire to inflict economic and political harm on the United States and its allies to the extent that we spend billions of dollars ramping up security systems and protective measures to overcome and deter any technological edge these cutting-edge systems could provide. This edge includes disruptive tactics aimed at major industrial and commercial operations, creative mechanisms to undermine and dilute the integrity and reliability of our own systems, cyber or nano-based probing gestures to penetrate and weaken erected defense systems, and the like.

The negative implications of the strategic hybrid threat on the scope and mission of homeland security will be profoundly complicated in the next decade. This will likely trigger defensive measures and system hardening steps to thwart cyber attacks of an increasingly clever and persistent nature. Some may be successful and some may not. Of course, our national defense agencies and the collaborative relationship between the U.S. Department of Homeland Security (DHS) and the Pentagon will also require recognition of an emerging threat, development of new doctrine to deal with it, and ultimately deployment of new systems and safeguards designed to offset the worst effects of mischievous technological inroads by our enemies. Without rigorous assessments of suspected offensive capabilities stemming from the extrapolation of advanced technologies there can be no sensible or appropriate defense. If some senior planners or strategists, for example, miscalculate and assume these technological breakthroughs will happen 25 years from now rather than in a few years then our overall preparedness for new threats will be poor.

Worse, there is an embedded bias among many senior leaders that rests on the unfortunately smug assumption that we already have enough sophisticated weaponry and hyper-technology in our arsenals that we need not be concerned about distant probabilities that enemies will acquire these new weapons any earlier than we expect. This may be true in some areas, but advances in neuroscience, synthetic biology, cybernetics, and nano materials make this a dangerous assumption. One additional idea that must be scrutinized is the extent to which advances in technology fundamentally alter the nature and variety of weapons of mass destruction (WMD). In fact, newer technologies may completely change our current grasp of weapons of mass destruction in ways we cannot contemplate. These new weapon systems may be equivalent to, or eclipse, the destructive power of some well-known WMDs.

Our need to revisit the homeland security threat spectrum specifically, and honestly assess how it may change over the next decade, is urgent. We should be explicitly examining today what the impact of advanced technologies may be in familiar areas where homeland security has the most at stake, such as:

- commercial aviation;
- maritime and sea trade operations;
- protecting and safeguarding vulnerable infrastructures; and
- altering the type and variety of weapons of mass destruction.

This wicked brew of threat issues, which brings Cold War, post-Cold War, emerging terror groups, post-9/11, and futuristic elements into alignment to complicate and exacerbate the future threat landscape, is the undeniable group of threat issues too often discounted or overlooked. They are the veritable black swans and unknown unknowns, which contain a mind-numbing array of unexpected, unanticipated, and unfathomed issues that can morph quite handily into threats we must address during the next decade. Preparations for the unexpected, for threats that exceed our current imagination, and for elements of strategic surprise cannot falter.

Finally, another important risk dimension for the future of homeland security is reckoning with persistent threats far into the next decade not likely to diminish. These include both novel and well-known threats such as radiological attacks, reverse-engineered drones, upticks in homegrown violent extremists (HVE), greater focus on soft targets, and low-tech forms of attack. In each case, developments in these areas must be watched closely.

Looking at the radiological threat first, we must realize that building a "dirty bomb" or a low-grade nuclear device does not require the utmost in technical sophistication. Missing, misrouted, or stolen radiological materials have happened at an impressive rate since 2005. So packaging high explosives together with a significant radioactive source like Cesium 123 or Strontium 90 represents no insurmountable hurdle. We know that even in a crude configuration, explosive release of radioactive material still yields a deadly effect that can render areas uninhabitable for several decades. Worldwide, more than 1,000 reactors operate nowadays, some producing electric power, others mostly used for research. Plutonium produced in either reactor type could be extracted for use in weapons. It is important to remember that not all nuclear facilities around the globe are as security conscious and careful to safeguard their facilities or oversee the management of radioactive waste as the United States. Challenges include how to secure the materials; how to detect, especially at a distance; how to render a potential device harmless; emergency response, cleanup, and public communication after a nuclear explosion; and determining who did it. All of these have engineering components; some are purely technical and others are systems challenges.[4]

Next, we must accept the fact that many advanced technologies can be reverse engineered or redirected to place our own modern high-tech systems in the hands of our foes. Drone technology is one such case. There exists no realistic governance policy on drone use inside the USA as a threat to aviation. Worse, drone technology is easily copied or borrowed anywhere on the globe, making possible terrorist or TCO use of drones for nefarious purposes.

Then there is the growth of HVE, which FBI and other law enforcement entities have been keenly aware of since 2007, but often powerless to stop. Our valued constitutional rights and protections must be balanced against "clear and present danger" to the extent that we tolerate morbid and externally inspired acts of a madman evolving into random shootings or related events unleashed with little or no warning. The challenge of HVE threats falls unfairly but predominately on local law enforcement, who seldom have the resources and wherewithal to thwart them. Again, we must prepare for an onslaught of low-tech attacks over the next decade against soft targets, including shopping malls, theaters, sports stadiums, train depots, and other conventional community gathering places where explosives, mass gunfire, or the use of crude chemical agents are a real possibility. Here the key message is that we are as vulnerable at the Iowa State Fair as we are in Times Square in the heart of New York City, and that no area, no place, is a safe haven and immune from terror. The overall aim of this venture is to erode and destroy the confidence of citizens in their government and in police forces to adequately protect them from harm.

In combination, looking at all of these threat situations arising together over the next decade may appear remote. That may be the case, but elements of TCOs, terror groups, and traditional nation-state enemies have time on their side. They will count on fatigue, lax security, weariness of special security measures, and related signs of intermittent vigilance as an opportunity worth exploiting.

Rounding Out the Strategic Hybrid Threat

To argue that this explicit combination of threat issues will dominate the next decade may be seen by experts as highly unlikely, however the last 60 years have contained enough proof that unexpected and unfathomed surprises of a "low probability-high consequence" nature can emerge to make global affairs even more disruptive and untidy. The "failure of imagination"[5] so often cited in the views of the esteemed 9/11 Commission does provoke a counter argument that we must allow for the distinct possibility of a new threat so unique and grave that its geopolitical impact can hardly be disputed.

Our collective imagination should allow for the possibility that an unexpected and unknown new threat could emerge during the next decade for which we have no strategy, no remedy, and no obvious solution. The scenario could entail a mix of terrorist and TCO activity that includes possession and deployment of unexpected weapons systems against familiar targets to a degree and with such ferocity that Western states find themselves the victim of strategic surprise. Often we even lack the ability to indulge futurist imagination by neglecting opportunities to examine unexpected threat scenarios in an explicitly unstructured manner using war games and expert Delphi methods. Experts have already warned against a "cyber Pearl Harbor" and some would speculate that unregulated advances in cybernetic, genomic, and robotic research during the next decade could unveil a weapons platform never seen before or anticipated.

Tracking the deterrence and "risks-of-failure" line of logic for a moment, we have hardly a comforting position to contemplate. Deterring actual use of weapons and malevolence from the strategic hybrid threat in future years is largely conditioned on at least three principles. First, the hybrid threat must be recognized for what it is and defensive doctrine and measures developed. Second, there is slim likelihood that prior conventional, or even asymmetric, articulated responses would be sufficient to deter a future author of hybrid threats, especially if the element of subterfuge or surprise were present. Third, the arguable advantages of launching a hybrid threat attack may be enough to compel its sponsors to push ahead regardless of risk because its effect may be devastating enough to attain more than a minimal victory. This contains the seeds of our downfall as we studiously cling to the well-known and reasonable while we tend to avoid idle speculation about an unarticulated future threat situation.

When we then posit the possibility that in facing these future threats, if not in combination, at least in parallel expression we find ourselves engaged in framing strategic responses that range from studied indifference to full-scale engagement and wholesale investment of blood and treasure. A summary analysis of future threats must take into account the following factors:

- traditional threats from classic rivals (Russia, China, etc.);
- threats from states of concern (Iran, North Korea, etc.);
- terror groups (Al Queda, ISIS, Boko Harum, etc.);
- transnational criminal organizations (Zetas, Kurosawa gangs, Russian mafia, etc.);
- dual-use advanced technology (nanoscience, cyber, genomics, etc.);
- low-tech, reverse-engineered, radiological attacks on soft targets;
- black swans (new unexpected nuclear state, new pandemic, etc.); and
- unknown unknowns (extreme drought, adverse downstream climate change, etc.).

The counterarguments against this tableau of threats include claims that the world will not be that much different ten years from now, that some of the threats identified will be more manageable, or that international objections and resistance are solid and sustained enough to thwart some of the worst possible outcomes. Some may also claim that combining emerging technologies to malevolent and deployable weapons systems is simply beyond the capacity of enemies, terrorists, or criminals in any event. These arguments also posit a low likelihood that all of the categorical threats listed above will emerge over the next decade in near simultaneous configuration.

Each nation will have to weigh its preparedness relative to others, and the USA more so because it still symbolizes a target of highest political value. Looking downstream a few years from now, we must also contemplate whether the subset of claims about the United States going it alone to fight and deter terrorism can be supported. Can we envision a threat or a set of circumstances where the USA would act unilaterally in its own defense or feel compelled to move ahead of our allies? History has shown it is possible.

Against all of these threats, or any combination thereof, most optimists would speculate that we would always enjoy the value and power of ad hoc or traditional alliances. They would argue strenuously that our allies have never abandoned us and that proven alliances like NATO would always come to our aid. Their primary point is that we would never encounter a threat that, because of its character or origins, as well as its indirect social, political, and economic costs, forces the United States to reckon with the hostile offender all alone. This is a brave but unpersuasive posture. Collectively viewed by like-minded states, the perceived strategic hybrid threat for the next decade surfaces significant security challenges—but only so long as most nations in the constellation of stable and secure states recognize and prioritize it as such.

The strategic hybrid threat is a new reality and a dangerous new challenge to those committed to robust homeland security, especially in the decade after 2015. At first, it seemed we were just adding the familiar threats emanating from TCOs and terror groups, but then we realized that advanced technology, the risks of black swans, and the delusional satisfaction nations derive when they discount the impact of unknown unknowns must be added to the threat matrix. Mixing criminal activity, nations of concern's misbehavior, and covert cooperative alliances with terrorist groups and the mean-spirited misdirection of advanced technologies means a terribly complex and convoluted threat will burden the USA for years to come. Is this a scenario that enjoys wide support? What factors make it less sinister than a resurgent Russia or a hegemonic China? Should we expect rival nations that have been quiet and peaceful for 20 years to remain so and decline to leverage a global threat situation where the USA is strategically distracted? Will such threat scenarios command continuing attention from our own strategic thinkers? Multinational criminal enterprises already influence economies and societies in Asia, Europe, and the United States. Is it far-fetched or irrational to imagine temporary alliances between TCOs and terror groups?

Hybrid groups that combine TCOs and terrorist organizations defy easy classification and routinely elude law enforcement and security measures meant to restrict their influence, which means the USA, and its likeminded allies, must find ways to dissolve its leaders, their funding, and support bases. There is also the power of social media and its thinly veiled invitation to youth seeking to make a difference and find expression for their own rage and frustration. What we often fail to see is a perfect storm, which sustains a constellation of threat issues all impinging on our security with the relentless rhythm of a 12 cylinder engine.

If the future threat is indeed novel, heinous, and far-reaching with its sharpest and most inventive weapons directed primarily at the United States, rather than anyone else, can we safely speculate that our historical allies will stand with us? Small-scale attacks on allies, friends, and trade partners may only dissuade them from confronting the threat for fear of lost revenue, eroded regional leverage, subsequent attacks, or simply to foster a fence-straddling position on more robust kinetic responses to their provocations. If this were to happen, we could find friends and allies aloof about formulating a robust form of retaliation. As a result, some alliances could be split. There are nations belonging to both the EU and NATO which may find themselves distanced from any posture we adopt that does not correlate with their own geopolitical calculus.

True enough, Article V of the NATO Treaty allows a victimized Alliance member to call upon the combined forces of the entire group of Alliance members, who are, in turn, obligated to render the support and assistance that they deem necessary. So most observers would envision a strong and determined NATO response against any threat targeting one of its alliance partners. NATO, through its allies, has special forces available to deal with terrorism against the Alliance, but what events would be the trigger for an all-out NATO response? In most cases, this response would entail all the trappings of a full-scale war with Alliance military resources aligned against the offending threat. In this case, it can be safely argued it is highly unlikely the United States would have to face its enemy alone. However, can we be assured of that regardless of the type of threat we face? Our nation's ability to provide leadership, to guarantee a level of global security, and to sustain an international partnership of like-minded nations is never guaranteed and is subject to tidal shifts in geopolitical relations. Will we have the same friends and allies in 2025 that we have now? Will our decisions and actions during the next decade result in wider global support for our positions or less? Can we surely contemplate any realistic threat scenario during the next decade where the United States must be willing to act alone?

Challenges for America's Homeland Security

In face of hybrid threat, one key issue that DHS must address is the degree to which the existing Threat Advisory System is working.[6] The DHS guide on the subject was issued in 2011 and seems to fall short of explaining the underlying rationale of its famous campaign, "If you see something—say something." The fault is not in the concept itself, but in failures to refine and clarify exactly what suspicious activity, anomalies, and strange events should be reported. To a sad degree, many in today's society see this task as one that falls on law enforcement and not on citizens in their daily lives. To increase the chances of having truly suspicious activities reported while safeguarding our treasured constitutional rights is no easy task. There must be a balance and there must be additional guidance from DHS detailing exactly what is "suspicious activity" and helpful information on how to spot such activity in the myriad situations which ordinary citizens confront each day. There is also a need to upgrade and refine the formal mechanism for suspicious activity reports (SARs), which local police departments are expected to draft and complete. Today, an unhelpful variety of approaches to this task exist among many of our cities, and this needs reform.

Finally, there are the immutable realities of fatal weaknesses stemming from failure to prepare for worst-case scenarios. In this instance, we are suggesting that a surprise attack, the neutralization of one of our prime technologies, or a strategically engineered probe designed to leverage our own technologies against us, must be seriously considered. There is no doubt we face a new spectrum of threats during the next full decade, ranging from the conventional and familiar to the exotic, novel, and unexpected, with a good deal of periodic insurrection, destabilization, and terrorism thrown in.

Despite our own failures, missteps, and errors in judgment over many decades, we have remained strong, stable, and secure. The "failure of imagination" cited by the 9/11 Commission members is a mistake we cannot afford to repeat.[7] Not without flaws, we expect that we have learned as much from 9/11 as we have from World War II, the Civil War, and our own societal revolutions. In the 300 years since 1750 that our nation discovered its own roots and purpose, can we remain confident that as we approach 2050, we have done all we could to buttress our own system and support those like-minded states that stand with us against the inroads of evil and chaos? Has the unfolding of human history for over 5,000 years thus far affirmed that observation? If so, then we have much less to fear of the future than we do of the past. However, in a world where the absence of geostrategic thinking can be fatal, is this a risk we are prepared to take? When we ponder a deadly mix of the old, the new, the familiar and the unexpected, and we remain supremely content that nothing can dislodge us from our perch, does that perspective give us a greater sense of security? What price do we pay and what pain must we endure if somehow we are mistaken?

Notes

1. Rebecca Harshbarger, "Terrorists Plotting to Blow Up 5 Planes in Christmas 'Spectacular': Report," *New York Post,* November 30, 2014, http://nypost.com/2014/11/30/terrorists-plotting-to-blow-up-5-planes-in-christmas-spectacular-report
2. See "Hybrid War," *Breaking Defense,* published February 15, 2013, http://breakingdefense.com/tag/hybrid-war
3. Sydney J. Freedberg, "Gen. Odierno: Budget Crunch Will Render Army Unready for Syria & Hybrid War," *Breaking Defense,* May 7, 2013, http://breakingdefense.com/2013/05/gen-odierno-budget-crunch-will-render-army-unready-for-syria-or-anywhere-else
4. "Prevent Nuclear Terror," National Academy of Engineering, last accessed February 25, 2015, http://www.engineeringchallenges.org/cms/8996/9134.aspx
5. *9/11 Commission Report,* 2014, 336, http://www.9-11commission.gov/report/911Report.pdf
6. "National Terrorism Advisory System," U.S. Department of Homeland Security, last published September 5, 2014, http://www.dhs.gov/national-terrorism-advisory-system
7. *9–11 Commission Report*, 336.

2

Border Management: International Experiences

NADAV MORAG

This chapter will provide the reader with a general sense of a few approaches taken internationally with respect to a number of aspects of border management. This is not intended to be an exhaustive cataloging of different approaches with respect to every facet of border management, rather the intent is to provide the reader with some vignettes of approaches employed by the European Union (and non-EU Schengen countries),[1] Australia, and Israel. Given the limitations of space, the chapter will be organized around three of the main components of border management with one example provided with respect to practices and strategies that touch on particular components.

Border Management: More than Just Border Security

Border management is a much broader concept than border security and generally includes four components:

- *border control*, that is, the regulation of access of people and goods to a country via land, maritime, and air borders and ports of entry;
- measures for *coping with immigrants, asylum claimants, and refugees;*
- measures for *enhancing security and coping with cross-border threats*, such as illegal drugs and other contraband, human trafficking, and terrorism; and
- efforts to *coordinate activities between and among domestic and foreign agencies* in order to enhance border control, the ability to provide asylum services, and the capacity to cope with cross-border threats.

It should also be borne in mind that many countries view border management as a layered activity that often begins far from the border or port of entry and can begin when a person applies for a visa or purchases an airline ticket.[2]

As noted earlier, this discussion will focus on the first three components and provide some unique examples.

Border Control

Border control involves the restriction of access for goods and persons to a country via its borders and ports of entry. The act of restricting access to a country's territory is a fundamental component of the exercise of national sovereignty and is ultimately necessary in order to ensure economic and social stability, as no country can afford to allow an unlimited flow of non-citizens and non-residents into its territory. Moreover, controlling access to a country is important in combatting cross-border crime and other security threats, and in regulating cross-border trade. Not surprisingly, countries with the capacity to effectively control their borders tend to jealously guard this right to limit and regulate access to their territory.

There is a strong desire on the part of governments to exercise sovereignty over their borders and national territories through, among other measures, effective border control. This however can have detrimental effects on trade, particularly in an era of globalization and just-in-time logistics, because effective border control often means longer wait times and more procedures, security checks, etc., for goods at ports of entry. As with the push and pull between civil liberties and security, generally speaking, the more border control the slower, and hence more expensive, the trade across borders and the more expeditious the trade, the laxer control of the border crossing or other port of entry. The key then is to find the right balance based on what is usually a pressing economic need to facilitate cross-border trade and the need for some level of efficient surveillance and control over people and goods entering the national territory. Given the need to expedite the movement of goods and persons and limited border control resources, in many cases, the degree of control exercised in a given part of the border (land or maritime) or port of entry will be variable and reflect the outcome of some sort of risk assessment. Naturally, no country can seal 100% of its borders 100% of the time and some countries operate under the expectation that they cannot even come close to truly sealing significant parts of their borders. This, essentially, is the approach taken by the United States with respect to sealing its southern border from illegal immigration.[3]

This tension between efficient cross-border trade and effective control of the border is evident, for example, in the U.S.-Canada border relationship. In the wake of the terrorist attacks of September 11, 2001, the United States concluded that it faced threats that required greater border control and that security measures would be tightened even at the expense of increasing the cost of cross-border trade. Increased control over U.S. borders with Canada caused considerable alarm on the Canadian side due to the fact that, in 2013, 75.8% of Canada's foreign trade was with the United States (by contrast, 16.4% of U.S. foreign trade is with Canada).[4] For Canada, which did not perceive itself to be, at least at the time, as threatened with international terrorism, facilitating quick and easy cross-border movement was more critical than more restrictive border control.

Europe represents an interesting attempt at finding a compromise between greater control of borders and the greater facilitation of cross-border traffic. The compromise involved the dissolution (except under very specific circumstances) of border controls between European countries and the enhancement of border controls on Europe's external borders.

The Schengen System and the European Union

The Schengen System (named after the town in Luxembourg in which the original agreement creating the system was signed) was implemented in its present form in 1995. The system began in 1985 as a plan to abolish border controls between France, Germany, Belgium, Luxembourg, and the Netherlands in order to promote trade, tourism, and facilitate the general movement of persons and goods across shared borders. In May 1999, the Schengen System was incorporated into the legal and institutional framework of the European Union following the signing of the Treaty of Amsterdam. This meant that most EU Member States became parties to Schengen and accordingly abolished border controls between themselves and other Schengen countries. Today, the Schengen area includes all Member States of the EU with the exception of Romania, Bulgaria, Croatia, and Cyprus (who are all interested in joining Schengen, but have not yet met its various criteria) and the United Kingdom and the Republic of Ireland, who have chosen to remain outside the Schengen area and thus maintain external border controls with Europe (Denmark, though part of Schengen, has retained the right to selectively apply some Schengen measures). The Schengen area also includes the non-EU countries of Switzerland, Norway, Iceland, and tiny Liechtenstein.

While Schengen countries thus were able to enjoy the benefits of free movement of persons and goods through the removal of border controls between EU Member States, this also had the potential to put them at great risk. The reason for this was that lax border controls in a Member State with a border or air or maritime links with non-Schengen countries could mean that criminals, terrorists, and/or large numbers of non-citizen foreign workers could access any part of the Schengen area. Consequently, the dismantling of internal borders also required the harmonizing of border controls at the Schengen area's external borders—that is, at airports, seaports, maritime borders, and land borders between Schengen signatory countries and non-Schengen countries or international waters. Those are purposefully referred to in the EU literature as *the external borders of the Member States* in order to stress the respect for national sovereignty.[5] Consequently, the *Schengen acquis* (the Schengen Agreement and associated rules and regulations) lays out specifics on common rules and procedures for admitting persons and goods into the Schengen area as well as coordination between Schengen signatories.

The *Schengen Borders Code*, promulgated by the European Parliament and Council of the European Union (council of Member States' ministers) in March 2006, created common standards for a common, Schengen-wide visa. Schengen signatories have agreed to supply visitors with a short-term visa (not exceeding three months in any six-month period) for travel anywhere within the Schengen area, provided that persons requesting such visas do not present a threat to public policy, internal security, public health, or the international relations of any Schengen members.[6] This means that if any Member State has flagged a given individual as a potential problem, that person will (if the system works properly) be denied entry at any access point to the Schengen area. In order for this to work, Schengen Member States needed to have a common database accessible at all ports of entry. This was accordingly created with the establishment of the Schengen Information System (SIS) database in 1995. Currently, the Schengen area has employed a second generation of the database (known as SIS II). It not only allows the authorities of Member States to issue alerts with respect to individuals posing a threat, but also provides alerts on missing persons (particularly children) and information on certain types of stolen or lost property including cash, vehicles, firearms, and identity documents.[7] The SIS II allows national law enforcement agencies to almost instantaneously share many types of data on suspected criminal activity. Additionally, the Schengen signatories within the EU allow each other's law enforcement agencies the right to engage in hot pursuit across mutual borders under certain conditions, and also grant each other a conditional right to engage in law enforcement surveillance activities in each other's territory.[8] A group of seven EU and Schengen signatories even signed an agreement in 2005, known as the Prüm Convention, which provides law enforcement agencies in each of the signatory countries direct online access to each other's law enforcement databases and enables them to exchange DNA, fingerprints, and other personal data.[9]

It should also be noted that national governments within the Schengen area still reserve the right to re-institute border controls, with certain restrictions and time limitations, if they decide they need to cope with a particular threat. For example, during the sixtieth anniversary of D-Day in June 2004 and after the terrorist bombings on the London Underground in July 2005, France reinstituted border controls for a brief period as a security measure. Portugal and Germany have done the same for sporting events such as the World Cup.[10]

In addition to operating a common database that is used at all Schengen area external borders, the EU Schengen countries with external land and maritime borders can also call upon mutual aid from other EU countries when the EU member with the external land or maritime border feels that its resources are being overwhelmed. This usually happens in the context of periods in which large numbers of refugees try to access EU territory via land borders (as in the case of the Greco-Turkish border or the two Spanish enclaves in Morocco) or maritime borders (such as the Mediterranean area, particularly near the Italian islands of Lampedusa and Pantelleria, which lay off the Tunisian coast, as well as the islands that comprise Malta). The need to provide resources to national border control authorities at the Schengen external borders led to the creation, in October 2004, of the European Agency for the Management of Operational Cooperation at the External Borders of the Member States of the European Union (thankfully known by the shorter appellation, FRONTEX). In addition to coordinating border control and security operations between Member States, FRONTEX also creates common training standards, trains national border protection forces, and carries out risk analyses and intelligence assessments. FRONTEX relies on the cooperation of national border police forces to provide it with operational capability and it coordinates joint operations, such as the surveillance of Mediterranean waters to target illegal immigration, and hence is dependent on the willingness of EU Member States to deploy their personnel in countries with external Schengen borders.

Since 2007, FRONTEX has been able to deploy specially trained teams of border guards from Member States known as RABITs (Rapid Border Intervention Teams). RABITs are designed to deploy for short periods of time under the command of the country's border security authorities in which they are deploying and they usually deploy alongside host-state border guards in order to provide greater legitimacy and legal cover for the teams (though they are authorized to carry weapons and use force if necessary).[11] Because of the temporary and crisis-mode deployment of RABITs and many other FRONTEX activities, it has been widely alleged that the Warsaw-based EU agency is essentially driven by political pressure employed by Member States with external borders on other Member States during crises rather than acting to provide a permanent solution to the fiscal and personnel pressures on border security agencies in states with external EU/ Schengen borders.[12]

While EU Member States have recognized the importance of having strong controls over Schengen external borders, the national governments of the Member States with external borders have thus far resisted the idea of giving up their ability to control their national border police forces in favor of creating a common European Border Guard as they—and other EU Member States—have no desire to confer law enforcement power on a supranational body.[13] At the same time, as with many other countries, EU Member States have outsourced external border enforcement to the private sector. Directive 2001/51 of the Council of the European Union requires that air and rail carriers check all travel documents and ensure that travelers have visas where needed, before they are permitted to travel. Air and rail companies are thus tasked with enforcing border security outside national territory and are fined if they do not comply. Moreover, Directive 2004/82 requires carriers to provide details about all passengers including names, nationalities, points of departure, etc., and carriers are fined for failure to comply.[14]

Schengen has undoubtedly provided tremendous benefits to Europe in terms of facilitating cross-border commerce and cooperation, but its desirability has been questioned when it comes to issues relating to security, refugees, and asylum. For example, on April 15, 2011, France and Italy became embroiled in a crisis that threatened continued French membership in Schengen. Faced with a large influx of some 25,000 Tunisian asylum seekers arriving by sea to the Italian island of Lampedusa, the Italian government became concerned that it did not have the resources to detain these individuals until the political refugees could be sorted out from the migrants who were simply searching for economic opportunities in Europe. This was despite the fact that under EU rules these people were technically Italy's responsibility because asylum seekers are supposed to remain in the territory of the first EU country they reach until that country's government can decide whether to accept them as legitimate refugees or to expel them because they do not fulfill the criteria of political, ethnic, and/or religious refugees facing persecution at home. The Italian government decided instead to provide these individuals with a temporary "humanitarian passport" effectively giving them access to the entire Schengen area. France, Belgium, and Germany protested the Italian action, fearing that many of these people, once freed in Italy and within the Schengen area, would illegally settle within existing Tunisian expatriate communities in their respective countries. Italy, of course, felt that it was expected to unfairly take on the fiscal and organizational burden of detaining 25,000 persons while other European countries benefited from not receiving an influx of these refugees that could strain their own social and institutional systems. Naturally, the French and others did not see it this way and argued that Italy was violating EU rules.[15]

Things eventually came to a head on April 15, 2011, when France temporarily re-established border controls with Italy and prevented trains from Italy crossing into French territory at Menton for six hours. Ultimately the crisis was resolved with the removal of French border controls, but this event illustrated the ongoing frustration on the part of the wealthier northern European Schengen countries with Schengen's working arrangements and a desire to rewrite Schengen rules (something that requires a qualified majority of Schengen Member States to agree to as well as the approval of the European Parliament) in order to strengthen external borders and allow more spot checks and other activities along Schengen internal borders.[16]

The EU has, from the outset, been characterized by tensions between the desire to bind Member States closer together, with the attendant economic, social, and political benefits, and the desire to maintain national sovereignty and autonomy. This debate is likely to continue, but the benefits of Schengen, in terms of the absence of tariffs, the movement of labor to where the jobs are, the facilitation of greater tourism, etc., are such that the EU will need to continue to strengthen external border control. The EU will also need to find the right organizational model that allows more burden-sharing between EU/Schengen countries that do not have external Schengen borders with those that do have such borders, while also safeguarding national sovereignty and governmental law enforcement and other legal powers.

Coping with Immigrants, Asylum Seekers, and Refugees

Coping with immigrants (legal and illegal), refugees, and asylum seekers that arrive at a port of entry or cross a land or maritime border is a critical part of border management. Generally speaking, an immigrant is someone who takes up domicile in another country legally or illegally, and who may or may not eventually become a citizen of their new home. Immigrants (whether legal or illegal) can choose to immigrate based on economic, lifestyle, family considerations, or other factors, but they are not considered to be individuals fleeing political or social persecution.

By contrast, an asylum seeker is, in effect, someone who is claiming refugee status, due to alleged persecution or the likelihood of persecution, but who has not yet been recognized as a refugee. A refugee is a person officially recognized by a country to be a refugee under the 1951 United Nations *Convention and Protocol Relating to the Status of Refugees*. This UN convention defines a refugee as someone who has a "well-founded fear of being persecuted for reasons of race, religion, nationality, membership of a particular social group or political opinion [...] and is unable or, owing to such fears, is unwilling to return [to his/her home country]."[17] It is, however, up to the state harboring the asylum seeker to determine if he/she meets the criteria of a refugee. In addition, even when granted this status, the host country still has the right to deport the refugee provided, under Article 33 of the *Convention and Protocol*, they are not forced to return to the country where they were persecuted or had reasonable fear of being persecuted (this is known as the principle of *non-refoulement*).[18] Consequently, the handling of asylum seekers and refugees primarily involves decisions made by the competent national authorities, though those authorities are expected to operate within UN guidelines and definitions.

Immigration

Developed democracies generally have an interest in encouraging immigration as a way of injecting new blood into the labor market, particularly with societies that have a rapidly aging population (which means a low birth rate) and that consequently recognize that they will need more young people to support burgeoning numbers of pensioners. Of course, encouraging large-scale immigration presents its own economic and social challenges, particularly with immigrant populations that do not possess, at least at the outset, the requisite education, linguistic, and/or vocational skills to succeed in the marketplace. Immigration also presents a challenge to national culture and social cohesion. Countries with a long history of encouraging immigration tend to view immigration with less angst and tend to be more successful in adapting their respective national cultures and social norms to integrate immigrants (who, in turn, tend to adopt majoritarian social values and ideas). But even countries that are comparatively new to modern immigration have found they need to learn to adapt to the reality, and need for, global population flows; Germany is a good example of this.[19] Of course, developed democracies also face the problem of illegal immigration, with numbers of illegal immigrants often being significant. For example, in the United States, there were an estimated 41 million immigrants in 2011, of which an estimated 12 million (just under one third) were considered to be in the country illegally.[20] In some cases, countries will turn a relative blind eye to illegal immigration, or otherwise focus more on illegal immigrants that pose a security threat (as does the United States) either because they informally recognize the economic benefits of undocumented immigrants or because they recognize that they can only prevent a fraction of people from the unauthorized crossing of borders.[21]

This is not the place for a full discussion of the benefits and costs of immigration or the impact immigration has on societies in the short, medium, and long term, but suffice to say that border management requires coping with and processing immigrants entering a country through a port of entry, as well as accessing its territory illegally via land, air, or maritime borders.

Asylum Seekers and Refugees

Asylum seekers represent a different challenge compared to both legal immigrants and illegal immigrants. As noted above, the treatment of refugees is ostensibly governed by international customary law. The UN *Universal Declaration of Human Rights* and the aforementioned *Convention and Protocol Relating to the Status of Refugees* recognize "the right of persons to seek asylum from persecution in other countries."[22] These UN guidelines act as the point of departure for asylum policies among developed democracies, though different countries will use different approaches and procedures in determining which individuals will have their asylum claims accepted and thus enjoy refugee status. Each country may also determine what these individuals may do in the host country while waiting for a determination as to whether they will be recognized as refugees.

Australia provides an interesting example of border management from the perspective of handling illegal immigrants and asylum applicants. As the country is a remote continent surrounded by water and also incorporates a number of islands, it has no external land border and its mainland is hard to access by sea (given that many illegal immigrants do not have access to large, sea-going vessels). Australia's geographic realities allow the country to more effectively control immigration and Australia, like most other developed democracies, caps the number of people per year who are eligible to apply for asylum status. For example, in the 2013–2014 fiscal year, the number of eligible asylum seekers was capped at 13,750—a fairly generous number as a percentage of the country's overall population (6%).[23] By way of contrast, the ceiling for refugee admissions in the

United States in fiscal year 2015 was 70,000 (2.21% of the overall population) and, in 2013, France took in 10,470 refugees (1.6% of its population), and even refugee-friendly Canada only took in 14,500 refugees (4.12% of its population).[24]

Australia is one of the pioneers of offshore resettlement for refugees, having, in many cases, effectively outsourced refugee resettlement to other countries. Australia's approach is to distinguish between individuals who make claims for asylum status when in the territory of another country and those who request asylum when already in Australia. Persons who arrive in Australia without a valid visa or other documentation can, under the Migration Act of 1951, be held in detention for as long as it takes to process their asylum claim and determine whether they should be granted refugee status or removed from the country. More recently, Australian authorities have released some detainees into the country while their claims are processed, but thousands more still remain in detention.[25] Australia also distinguishes between refugees who arrive by air (who are processed in Australia) and those who arrive by sea. Since August 2012, refugees arriving by boat are processed offshore, either on one of Australia's smaller islands (such as Christmas Island, an Australian territory close to Indonesia), or in third countries such as Papua New Guinea and Nauru—known as "Regional Processing Countries." Regional Processing Countries have agreed to take in persons arriving to Australia by sea and, if they are deemed to meet the criteria of refugees, are resettled in those Regional Processing Countries or other participating Pacific Islands countries.[26]

The Australian government's policy of processing asylum-seekers offshore and in third countries, commonly known as the "Pacific Solution," includes three components. Firstly, the government determined that certain offshore Australian territories (Christmas Island, Cocos Island, and Ashmore Reef) would be excluded from Australia's migration zone and consequently asylum seekers would be unable to apply to Australia for refugee status from these locations. Secondly, the Australian Navy was authorized to interdict "boat people" and prevent them from reaching the mainland. Finally, as previously noted, Nauru and Papua New Guinea agree to become Regional Processing Countries and detain and process asylum seekers.[27] More recently, Australia added Cambodia to the list of Regional Processing Countries despite criticism that it, as well as Nauru and Papua New Guinea, do not have the infrastructure or the capacity in their respective justice systems to adequately determine who should be awarded refugee status.[28] Criticism of the policy led to its temporary suspension in 2007, but it was ultimately perceived as providing an effective solution, from the Australian perspective, that served to deter asylum seekers and consequently it was revived in 2012.

This process of shifting the burden of dealing with refugees to offshore territories with limited infrastructure or to poor third countries has been controversial, but will likely continue because, for the most part, it keeps boat people away from mainland Australia. In supporting the effective reestablishment of the "Pacific Solution," then Prime Minister Julia Gillard argued that it saved lives because it deterred people from setting out for Australia and thus risking their lives at sea.[29]

Securing the Border and Coping with Threats

The third component of border management has to do with physically securing the border and ports of entry in order to thwart the smuggling of drugs, weapons, and other contraband; the cross-border trafficking of human beings; and the infiltration of terrorists or other people that threaten public safety. This is an issue that concerns essentially every country, though the degree to which borders and ports of entry are secured can vary from the complete absence of any border barriers—as between many developing countries, which usually lack the resources to create effective barriers, or in remote regions across the globe—to heavily fortified barriers supplemented by patrols, cameras, and other tools.

Israel's borders with its neighbors, including the Israeli Separation Barrier in the West Bank, provide good examples of heavily fortified borders designed to create a complete denial of access, not only to terrorists, but to contraband smugglers of various types. Construction on Israel's Separation Barrier with the West Bank began in 2002 in the wake of what proved to be the most sustained and bloody terrorism campaign the country had ever faced (in the context of the Second Palestinian *Intifadah*). Some 1,200 Israelis were killed and over 8,000 injured in terrorism attacks and approximately two-thirds of those attacks consisted of suicide bombing operations. As virtually all of the suicide bombers attacking Israeli urban populations came from the West Bank, and as it proved impossible to prevent people from accessing Israeli territory from the West Bank due to the absence of any barriers to pedestrian traffic outside the road network, it was decided in 2002 to construct a physical barrier. That barrier consists primarily (over 90%) of a multilayered fence system and in urban areas or where the barrier abuts major roadways, the fence gives way to a high concrete wall designed to prevent sniper fire.

The security fence consists of a series of obstacles. The West Bank side of the barrier includes a pyramid-shaped stack of six coils of barbed wire, then a barbed wire fence, followed by a ditch (to prevent vehicles trying to break through the barrier). This is followed by a road for Israeli military patrols, then a high chain link and barbed wired fence with an intrusion detection device that notifies the local army command post when someone or something has touched the fence. One the western side of the fence is a sand road that is smoothed over regularly, so that if someone crosses the fence this will leave footprints. Next comes a road for patrol vehicles, and this is followed by another sand road to detect intrusions. Finally, at the very western edge of the barrier, are more rows of barbed wire coils and another barbed wire fence.[30] The barrier is augmented with electronic measures such as motion-sensor detectors, excavation detectors, and cameras. The barrier is patrolled by units of the Israeli military as well as Israel's Border Police (a paramilitary gendarmerie-like force that, when conducting these patrol missions, operates under the military's operational command). The construction of the barrier is credited with the reduction of suicide bombing from the West Bank so that the number of suicide bombings went from a high of 53 in 2002 (pre-barrier) to zero in 2009.[31]

On the border with Egypt, Israel has constructed a different type of barrier that reflects the contrasting set of challenges that it faces on its southwestern edge. Since the signing of the Israeli-Egyptian peace treaty in 1979 and the final pull-back of Israeli forces to the present border in 1982, Israel's border with Egypt has arguably been the country's calmest border. Infiltrations still occurred on a regular basis from the Sinai, but these were of a criminal nature involving the smuggling of drugs and human trafficking, rather than involving terrorism threats. The border barrier, if it can be called that, was, in most places, a low barbed wire fence barely a meter high that presented no serious obstacle. As Israel traditionally pays much more attention to security threats involving terrorism and war as opposed to those related to criminal activity, the smuggling problem on the border was not seen as justifying the huge expenditure required to fortify the 240 km-long border. Official attitudes began to change toward the end of the 2000s as Israel faced a massive influx of persons claiming refugee status from Africa (primarily Eritrea, Sudan's Darfur region, and South Sudan, but also from the Ivory Coast and Nigeria), who were traveling through Egypt (despite the risks of being shot at by Egyptian border guards and the hardships of travelling through the harsh Sinai desert). As a percentage of its population, in 2011 Israel had to cope with ten times as many illegal immigrants as France, twenty times as many as Italy, and one hundred times as many as Spain.[32] This influx of persons led to the usual phenomena associated with large-scale migration: migrant ghettos appearing in large cities, an increase in petty crime, as well as fears of potential radicalization and recruitment of migrants by terrorist groups. Since the primary objective of border control in the south was to prevent illegal immigration, the barrier there had to consist of a very high fence (five meters) to make it difficult for people to climb over it. At the same time, given that there were fewer concerns regarding terrorist infiltration, it was not seen as necessary to create as much of a layered barrier as in the case of the fence in the West Bank. Nevertheless, the fence was constructed in such a fashion that its eastern side is covered with razor wire and just to the east of the fence are stacked rolls of barbed wire arranged similarly to those used for the West Bank Separation Barrier in order to more effectively deter smugglers and potential terrorists, as well as illegal immigrants.

Over time, the need for the fence to also serve as a barrier to terrorist infiltration increased as Egyptian rule over the Sinai became even more tenuous with the tottering and then collapse of the Mubarak regime and its replacement by the Morsi government. During this period, global Jihadi groups, as well as Palestinian groups such as Hamas, began to view the Sinai as a useful launching point for attacks into Israel. The real wakeup call in this context came on August 18, 2011 when a platoon-sized force of Jihadis broken into separate squads, infiltrated Israel via the Sinai, attacked civilians in a bus, and engaged the Israeli military in pitched battles. The level of coordination and skill involved in the attack was such that it took the Israeli military by surprise and led it to start viewing the fence as a critical counterterrorism tool, in addition to its role as a deterrent to illegal immigrants.[33] The fence was also designed to be strong enough to withstand being run into by a high-powered vehicle. In early August 2012, a Jihadist terrorist attack on an Egyptian military base resulted in the hijacking of a jeep and armored car, which were then driven toward the Israeli side of a border crossing and were able to penetrate the fence around the border crossing (this was not the new fence being constructed). This made it even clearer to the Israeli authorities that the fence would need to be able to withstand significant impact. Since its construction, the fence has significantly reduced access from Egypt and even led to the rise in Israel of the price of marijuana since the fence has cut down the previously extensive smuggling of the drug into Israel from the Sinai.[34]

The examples of the Separation Barrier in the West Bank and the Israel-Egypt border suggest that it is possible to secure borders to a significant degree, albeit at a high financial cost and often requiring a major footprint in terms of the width of land taken up by various barriers. While the threats facing other countries may not require as extensive a set of physical barriers, the combination of physical barriers, alarms, cameras, and other technologies, have proven effective and this methodology, perhaps in modified form, may be used to enhance the security of the border and the capacity to reduce terrorism and criminal threats.

Conclusion

As can be noted from the brief discussion above, border management involves a wide array of activities designed to both facilitate the movement of goods and persons across borders as well as restrict that movement in cases where it may negatively impact the economy and/or social cohesion, or where it may increase criminality or terrorism. The examples provided above are, in some ways, cases that provide maximalist options for dealing with various border management requirements and challenges. The Schengen area and the EU Schengen arrangements provide what is arguably the maximalist option in terms of facilitating movement across borders (that is, internal borders) coupled with an attempt (of uneven success at present) to strengthen external borders. Consequently, Schengen is probably the clearest example of unencumbered movement across national borders in the quest to facilitate trade and the free movement of people and thus enhance economic potential. At the other end of the spectrum, Australia represents a somewhat maximalist approach to keeping out unwanted immigrants and asylum seekers (as well as shifting the burden of coping with legitimate refugees, at least in part, to third parties). Finally, Israel represents a maximalist perspective in terms of physically securing borders via a combination of physical barriers and technology. Ultimately, each country will need to decide for itself where it falls along the continuum of facilitation of movement and trade on the one end, or the maximization of security and control of borders and ports of entry on the other.

Notes

1. See also Michel F. Bosco, "Trends on Security Research in Europe," Chapter 12 in this volume, 211–212.
2. Australian Government, Department of Immigration and Multicultural Indigenous Affairs, *Managing the Border: Immigration Compliance*, 2004–05 Edition (Canberra: Commonwealth of Australia, 2005), 3.
3. "Napolitano: 'You Cannot Seal a Border,'" *CBS News*, May 22, 2014, http://houston.cbslocal.com/2014/05/22/napolitano-you-cannot-seal-a-border
4. Industry Canada, *Canadian Industry Statistics* (Ottawa: Government of Canada, 2013), http://www.ic.gc.ca/eic/site/cis-sic.nsf/eng/h_00029.html; United States Census Bureau, *Top Trading Partners—December 2013* (Washington, D.C.: U.S. Census Bureau, 2013), https://www.census.gov/foreign-trade/statistics/highlights/top/top1312yr.html

5. Sergio Carrera, *The EU Border Management Strategy; FRONTEX and the Challenges of Irregular Immigration in the Canary Islands,* CEPS Working Document, no. 261 (Brussels: Center for European Policy Studies, 2007), 9.
6. Regulation (EC) No. 562/2006 of the European Parliament and of the Council of 15 March 2006 Establishing a Community Code on the Rules Governing the Movement of Persons across Borders (Schengen Borders Code), *Official Journal of the European Union,* L 105, April 13, 2006, http://eur-lex.europa.eu/LexUriServ/LexUriServ.do?uri=OJ:L:2006:105:0001:0032:EN:PDF
7. European Commission, "Questions and Answers: Schengen Information System (SIS II)," press memo, April 9, 2013, http://europa.eu/rapid/press-release_MEMO-13-309_en.htm
8. Matti Joutsen, *The European Union and Cooperation in Criminal Matters: The Search for Balance,* HEUNI Paper, no. 25 (Helsinki: The European Institute for Crime Prevention and Control, 2006), 11–12.
9. Council of the European Union, *Prüm Convention,* July 7, 2005, http://ec.europa.eu/anti_fraud/documents/data-protection/dpo/prumtr.pdf
10. "Q&A: Schengen Agreement," *BBC News,* March 4, 2013, http://www.bbc.com/news/world-europe-13194723
11. Sergio Carrera and Elspeth Guild, *Joint Operation RABIT 2010—FRONTEX Assistance to Greece's Border with Turkey: Revealing the Deficiencies of Europe's Dublin Asylum System* (Brussels Center for European Policy Studies, 2010), 1–4.
12. Carrera, *EU Border Management Strategy,* 12.
13. Jorg Monar, "The Project of a European Border Guard: Origins, Models and Prospects in the Context of the EU's Integrated External Border Management," in *Borders and Security Governance,* ed. Marina Caparini and Otwin Marenin (Geneva: Geneva Centre for the Democratic Control of Armed Forces, 2006), 5.
14. Elspeth Guild, *Danger—Borders under Construction: Assessing the First Five Years of Border Policy in an Area of Freedom, Security, and Justice* (unpublished), last accessed February 24, 2015, http://www.lse.ac.uk/europeanInstitute/research/hellenicObservatory/pdf/Various/ElspethPAPER.pdf
15. John Lichfield, "EU Border Deal under Threat in Row Over Italy's Migrant Burden," *The Independent,* April 19, 2011, http://www.independent.co.uk/news/world/europe/eu-border-deal-under-threat-in-row-over-italys-migrant-burden-2269641.html
16. Hugo Brady, "The Schengen Crisis in the Framework of the Arab Spring," *Panorama* (2012), http://www.cer.org.uk/sites/default/files/schengen_arab_Spring_Dec12_IEMED.pdf
17. United Nations High Commissioner for Refugees, *Convention and Protocol Relating to the Status of Refugees* (Geneva: UNHCR, 1951), 14, http://www.unhcr.org/protect/PROTECTION/3b66c2aa10.pdf
18. Ibid., 28.

19. Jenny Hill, "Germany Struggles to Adapt to Immigrant Influx," *BBC News*, November 2, 2014, http://www.bbc.com/news/world-europe-29686248
20. Chiamaka Nwosu, Jeanne Batalova, and Gregory Auclair, *Frequently Requested Statistics on Immigrants and Immigration in the United States* (Washington, D.C.: Migration Policy Institute, 2014), http://www.migrationpolicy.org/article/frequently-requested-statistics-immigrants-and-immigration-united-states
21. H. A. Goodman, "Illegal Immigrants Benefit the US Economy," *The Hill*, April 23, 2014, http://www.thehill.com/blogs/congress-blog/foreign-policy.203984-illegal-immigrants-benefit-the-us-economy
22. United Nations High Commissioner for Refugees, *Convention and Protocol Relating to the Status of Refugees*, 2.
23. Australian Human Rights Commission, *Asylum Seekers and Refugees Guide* (Canberra: AHRC, n.d.), last accessed February 24, 2015, https://www.humanrights.gov.au/asylum-seekers-and-refugees-guide#who
24. Immigration Policy Center, *Refugees: A Fact Sheet* (Washington, D.C.: IPC, 2014), http://www.immigrationpolicy.org/just-facts/refugees-fact-sheet; "France Rejects Four out of Five Asylum Seekers," *The Local FR*, March 24, 2014, http://www.thelocal.fr/20140324/france-asylum-refugee-rejects-four-out-of-five-asylum-seekers; Immigrant Settlement Services of British Columbia, *Refugees: The Big Picture* (Vancouver: ISS of BC, n.d.), last accessed February 24, 2015, http://issbc.org/prim-corp-nav/our-work-with-refugees/Refugees-big-picture
25. Australian Human Rights Commission, *Asylum Seekers and Refugees Guide.*
26. National Legislative Bodies/National Authorities, *Regional Resettlement Arrangement Between Australia and Papua New Guinea*, July 19, 2013, http://www.refworld.org/docid/51f61a504.html
27. Jaffa McKenzie and Reza Hasmath, "Deterring the 'Boat People': Explaining the Australian Government's People Swap Response to Asylum Seekers," *Australian Journal of Political Science* 48, no. 4 (2013), 418.
28. Ian L. Neubauer, "Australia's Plan to Outsource its Refugee Problem to Cambodia Won't Work," *Time*, September 29, 2014, http://www.time.com/3443252/australia-asylum-seeker-refugee-cambodia
29. Adam Gartrell, "Teams to Give Verdict on Island Facilities," *Sydney Morning Herald*, April 20, 2012, http://www.smh.com.au/national/teams-to-give-verdict-on-island-facilities-20120819-24gr1.html
30. Israel Ministry of Defense, *Israel's Security Fence Operational Concept* (Tel Aviv: MOD, January 2007), http://www.securityfence.mod.gov.il/Pages/ENG/operational.htm
31. Ido Rosenzweig and Yuval Shany, "A Decade of Palestinian Terrorism—Report by the Israeli Security Agency," *Terrorism and Democracy*, no. 14 (n.d.), last accessed February 24, 2015, http://www.en.idi.org.il/analysis/terrorism-and-democracy/issue-no-14/a-decade-of-palestinian-terrorism-%E2%80%93-report-by-the-israeli-security-agency

32. Yonatan Jakubowicz, "Illegal Infiltration to Israel: It's the Economy Stupid!" *Jerusalem Post*, January 7, 2014, http://www.jpost.com/Opinion/Op-Ed-Contributors/Illegal-infiltration-to-Israel-Its-the-economy-stupid-337406
33. Amos Harel, "On Israel-Egypt Border, Best Defense Is a Good Fence," *Ha'aretz*, November 13, 2011, http://www.haaretz.com/print-edition/features/on-israel-egypt-border-best-defense-is-a-good-fence-1.395239
34. Maayana Miskin, "August: Zero Illegal Infiltrations along Israel-Egypt Border," *Arutz Sheva*, September 1, 2013, http://www.israelnationalnews.com/News/News.aspx/171488#.VKh2LyvF98E; Maayan Lubell, "Israel Weed Smokers Choked by Border Fence with Egypt," *Reuters*, January 30, 2014, http://www.reuters.com/article/2014/01/30/us-israel-drugs-idUSBREA0T0MB20140130

3

The Three Mile Island Nuclear Disaster from an Emergency Management Perspective

KEVIN J. MOLLOY

Introduction

Incident Command System (ICS), Emergency Support Function (ESF), and Emergency Management Assistance Compact (EMAC) are familiar lexicon in today's world of emergency management. While those particular terms were not in existence or utilized often in the 1970s, they were all, in fact, utilized during the response to the Three Mile Island accident.

From an emergency management perspective, the Three Mile Island (TMI) accident of March 28, 1979 and the attacks on America in September 2001 bear similar characteristics. As an individual involved personally in both events, I will endeavor to provide you with food for thought as you continue your learning progression in this ever-expanding area of emergency management. Incorporating my actual personal experience takes this chapter out of the pure textbook realm. It is an attempt to meld the application of emergency management principles and practices with academics. This should afford you the ability to view both the practicality of the application of what you learn in an academic environment and what, in fact, may really occur during a man-made, natural, or technological emergency.

I will enumerate the number of lessons learned that enable emergency management to function more effectively in this day and age. I must advise you that in over 40 years in emergency management, I have learned that disasters and emergencies do neither perfectly follow written plans nor any textbook.

This is a critical point to remember: In the years preceding the Three Mile Island (TMI) accident, the United States civil defense effort and programs were still centered on either stopping and/or surviving a nuclear attack. The Cold War was not quite over. If, as a result of planning efforts on how to save people from the (atomic) "bomb," additional planning emerged dealing with preparedness and response to natural or man-made events, well, that was a bonus.

At the time of the TMI accident, there was no Federal Emergency Management Agency (FEMA). FEMA came into formal existence on April 1, 1979. There existed the Defense Civil Preparedness Agency (DCPA). All of the DCPA program requirements, monies provided to state, county, and local organizations for personnel, training, and facilities were based on the concept of how will these dollars be used from a civil defense perspective. Funds from the DCPA were matched by the county at a 50–50 ratio. From the 1950s through the 1970s, not a great deal of money was spent on what we view as current emergency management planning or training programs and practices. Ongoing in many Pennsylvania counties were efforts to utilize other sources of funding such as the Law Enforcement Assistance Agency (LEAA) to upgrade and enhance communications equipment installed and operated in county and city Emergency Operating Centers (EOCs) or Communications Centers.

In the years preceding the terrorist attacks of September 11, 2001, there was no U.S. Department of Homeland Security (DHS). Following September 11, DHS was established with over 20 individual government agencies becoming integral members of the new organization. Until the formation of DHS, FEMA was the federal agency tasked with providing funding and other support to state and local governments. Most emergency management programs were based on preparing and planning for and utilizing an all-hazards approach at all levels of government. So at the time, terrorism or homeland security planning was not necessarily on the front burner, particularly at the state and local levels of government. Additionally, in the 1970s at the higher level of government, the independent development of "new" programs occurred. These new programs often were developed with little, if any, coordination with state and local governments or the citizens. Thus, you perhaps are now seeing some similarities between TMI and the September 11, 2001 events: They included but are not limited to organizational structure and programmatic areas and response abilities. The issue of a lack of coordination between levels of government and various departments, agencies, and disciplines is not unique to America. During my years with DHS-FEMA, I taught various emergency management courses in a number of foreign countries and saw this issue occur many times.

Immediately following the TMI accident, the prime focus of spending monies for planning, training, exercises, and equipment in America was centered on response to nuclear accidents. Immediately following the events of

September 11, 2001, planning, training, and expenditure of funds at all levels of government focused on terrorist emergencies. In the ensuing years between 1979 and 2001, the system had not changed. Approaches in emergency management remained focused on the disaster "du jour" mentality as opposed to disaster preparedness based on risk assessment leading toward planning, training and exercises, and response equipment for any type of hazards and incorporating lessons learned from response to either man-made, natural, or technological disasters. It is at this juncture that I wish to provide a perspective on emergency management as it relates to the Three Mile Island accident of March 28, 1979.

A County Perspective on the Three Mile Island Accident

My appointment as Dauphin County Pennsylvania Civil Defense Director and County Communications Director in 1973 was approved by Governor Milton Shapp at the recommendation of the Dauphin County Commissioners. I was the youngest of the 67 County Directors at the time. The significant age difference resulted in my vision of what methods and goals the County Agency should be utilizing for planning and training differing significantly from my counterparts across the Commonwealth. My focus was not on civil defense, with its emphasis on bomb shelters. My main priorities were strengthening emergency planning, overseeing the installation of a new, countywide communications system, and the institution of the 911 telephone system. Upon reviewing Dauphin County history chronicles relating to emergency management, I learned of many major events/emergencies, such as the 1972 Agnes Flood, which killed a large number of people, to several hazardous materials events resulting in evacuations, airplane crashes, and train derailments. Thus, in Dauphin County, based on history, the probability of a flood, hazardous materials incident, or other emergency event occurring was more likely than a major accident at Three Mile Island.

However, I did recognize that much of the new emergency planning approach we were utilizing would in fact be beneficial if we ever had to deal with an issue at TMI. This portion of the chapter relates and deals with my personal recollections of actions taken during the course of the TMI accident. The information is taken from notes that I made both at the time of the accident, as well as communications tapes and logs from the Dauphin County Emergency Operations Center.

Wednesday, March 28, 1979

March 28, 1979 started off like most days. I woke up, shaved, showered, dressed, and made coffee. I woke up my children for school and my wife Carol for work. At 7:05 a.m., I was in the back of the house, when the telephone rang. My daughter shouted, "It's Margaret Reilly for you!" I was surprised. I thought it strange that she would call me at this hour and had a sinking feeling in my stomach that something must have happened at Three Mile Island. Maggie Reilly, at the time, was with the Pennsylvania Department of Environmental Resources, Bureau of Radiological Protection (BRP). She and I had worked together on occasion dealing with testimony on the licensing of Unit 2 of TMI regarding off-site emergency response. In the past, we had done telephone notification drills—yet this call seemed different.

"Good morning, let me ruin your day," she said to me. Ms. Reilly then proceeded to tell me in very non-technical terms that there is a problem at the plant. My first thought was do we have to evacuate. She responded, no. Ms. Reilly continued on, saying at this time there was a problem with some equipment at the plant, that there is some "hot s***" in the containment building, but reiterated that no evacuation was necessary at this time. In her own unique way, she advised me to head into work a little early. I told her I would head in immediately. As I pulled out of my driveway, I noticed something strange on this cloudless sunny morning. There was no condensation steam plume at TMI. While not totally understanding the significance of what I was not seeing, it made me more aware of a problem at the plant.

While driving the 13 miles to Harrisburg, in my mind I am already planning my initial course of action. First, verify with the Pennsylvania Emergency Management Agency (PEMA) what the latest situation information at the plant is. Next, I will call the local Civil Defense Directors and other emergency personnel in the political subdivisions closest to the Three Mile Island plant including Londonderry Township, Middletown Borough, Lower Swatara Township, Royalton, and Highspire Borough. They all are in Dauphin County. Once at the office, I checked in with the dispatchers on duty in the communications center. Because of the direct call from the TMI control room and my early arrival at work, they seemed confused but not overly concerned. Again, consistent with emergency management practices, I prioritized my course of action. While we did not use the Incident Command System at the time, it turns out that is exactly the procedure I initially utilized for operation of the EOC. Eventually, I also set up my volunteer

EOC staff by function (law enforcement, medical, etc.), which is exemplified by the 15 ESFs utilized in emergency management today.

I immediately called the state (PEMA) to confirm what they knew to ensure our information was the same. I had the base dispatchers broadcast over the police and fire network for various chiefs to call in as soon as possible. Within a matter of a few minutes, we had briefed all of the emergency management, fire, and police departments of communities near that facility. I also advised the individuals that I had briefed to pass on the information to elected officials. Elected officials have the legal responsibility for the health, safety, and welfare of the residents of the community. At that time, emergency planning was not a major priority with most communities in Dauphin County. Potholes and other issues were on the front burner of priorities of elected officials. Based on this initial notification from my office and responses from local personnel, I felt a level of not only deep concern, but an additional feeling of "What do I do now?" developing down at the local level.

At 7:37 a.m., a dispatcher in the communications center received a call from Dick Dubiel, Senior Health Physicist, advising of a radiation emergency at TMI. My dispatcher yelled to me, "Kevin, it's Three Mile Island." Picking up the phone, I hear "Kevin, it's Dick Dubiel." I respond "Yes, Dick." He said, "OK, we're in for real," to which I simply responded, "OK." Dubiel went on: "What type we're not exactly sure. We got the core covered right now. I don't think we got a real big problem, but we've got some, ahh, bad radiation readings that, in fact, could be erroneous, but we can't rely on that—OK?" Dubiel concluded his call by saying, "What I need is, I need to get Maggie Reilly informed and back in touch with us as soon as possible, OK?" I then contacted Ms. Reilly, who advised me that a representative from her agency was on site and monitoring events. This request from Dubiel to "immediately" talk with Ms. Reilly certainly added a sense of urgency and more unknowns to the situation. I then received a call from Mike Pintek of WKBO radio asking what was going on. I told him simply that there was a problem at Three Mile Island and that he should call PEMA for specific information.

At the time, I believed informative public information should come from primarily one source. In today's emergency management, the existence of Joint Information Centers (JICs) and pre-prepared emergency information has been proved to be the most effective means of providing the most accurate and comprehensive information to the public. During the TMI accident response, the lack of a clear communications chain of command combined with incorrect information was a definite hindrance to easing public concerns. This shortcoming of not operating in a coordinated fashion did not allow for the use of what we now refer to as the *common operating picture.* An accurate common operating picture is essential to both responders and citizens. This communications failure would haunt everyone in the days to come.

Throughout the rest of the day, there were isolated requests from different organizations regarding events at the crippled nuclear plant. We received reports of recreational boaters gathering in the vicinity of the island and small aircraft circling the cooling towers, and if we could we do something about that. We dealt with each request. During this period, the main word from the state emergency management agency is things are stable, keep working on plans, and stay cool. My staff and I maintained contact with and passed information to the local emergency management people throughout the afternoon and evening. Day 1 ended with frayed nerves on the part of a great deal of people. Most of the world still had never heard of Three Mile Island. Most of the people around the plant slept, although some better than others. I stayed overnight in the emergency operations center. That night, I kept watch over the state teletype and I also reviewed existing plans and procedures. Based on my interface with staff from the Bureau of Radiological Protection, I possessed a limited knowledge of the operations of a nuclear power plant and a basic understanding of the extreme danger of radiation. I was under the impression that any accident at the plant would, in all likelihood, be confined to the plant itself with no harm to the public—a lesson I would need to unlearn as a result of the TMI accident.

Thursday, March 29, 1979

During the night, I receive updates from the state. PEMA advised that "that the Reactor should be in the desired heat and pressure range with the hour." Once again, we pass that information along the response chain. This overnight shift was abnormally quiet. I assume all of the drunks, burglars, and thieves know something we do not and are taking the night off. I chuckle as I make this observation to my dispatchers—we now have a new crime prevention program.

Sitting in my office in the basement of the courthouse, I finally get a chance to review the newspapers. Local print media is dealing with the event in a rational manner. I did get reports from relatives, residents, and responders that some of the out-of-town and out-of-state newspapers are sensationalizing the accident. At times, it made me wonder if I am involved in the same emergency they are airing or printing stories about. I go back and review some of my hand written notes regarding the condition of the plant, including such comments as

"10 millirem (mR) reading in Royalton." Additionally, the readings at Harrisburg International Airport, located about two miles from the plant, were reported as 2 mR/hour. I also see readings of .10 mR/hour and .01 mR/hour at Middletown Borough and Hummelstown Borough, respectively. I was wondering what the readings were in Londonderry Township where the plant and my home were actually located. I must confess, the words "aircraft monitoring" and "plume" were of more interest to me than specific readings at this point. In reviewing our actions to date, I felt confident that we dealt with the TMI problem in a reasonably effective fashion, with one exception: I hadn't personally notified Royalton Borough. We accomplished notifications as planned and reviewed other specific preparedness actions, communicated with nearby counties that might provide resources, and reviewed our evacuation procedures if they became necessary.

To this point, TMI certainly doesn't appear to be one of the all-time biggies, but what the heck, you can't tell in this business. The ability to actually determine "how big was big and how bad was bad" was lacking due to the uncertainty, which precluded our ability to have a clear common operating picture. By this point in time, it was evident that unfolding technological and political crisis factors were having a negative influence on our dealings with local elected officials, responders, and the public. Notes that I made at the time show monitoring was being done by both the Nuclear Regulatory Commission (NRC) and the State Bureau of Radiological Protection and that we were on alert until further notice. The NRC said no danger to the public, no core damage, and they commend the Metropolitan Edison (Met Ed) utility, which owned Three Mile Island, for professionalism. I also jotted down a comment that reads "Ventilating cycle, slight contamination throughout the system—through towers." I thought that TMI staff on site had all well in hand.

March 29, 1979: Morning

At 5:45 a.m., State Civil Defense says the reactor is stable. Around 6:30 a.m., I make my first round of telephone contact to the local Civil Defense Directors near the plant. My Deputy Director, secretary and other staff arrive, dispatchers change shifts, and the courthouse comes alive. Most departments operate more or less in a business as usual mode.

PEMA sends its latest situation report ("Sitrep") via teletype around 9:00 a.m., covering the period of inception of the accident through 7:00 a.m. that morning. Off-site radiation levels very low, no one outside the plant exposed to more than 100 mR/hour. The facility (TMI) may revert to ventilation. Situation appears more secure. Continue readiness posture. The courthouse reporter for the Harrisburg newspaper stops in to talk with me. We discuss the situation and what actions we have taken and intend to take to respond to the TMI accident.

I learned the difference between local, national, and international news media functioning in major events or emergencies. Locals calmly and professionally report what they can confirm, while some out-of-town media behave in a questionable manner. I received a report from the field concerning a film crew in an unmarked van pulling in front of a house, sticking a for-sale sign in the lawn, and filming the scene. They then removed the sign and sped off. I never tried to verify that report, but when I was interviewed by the media, I always said, "Listen to people you can trust." Meanwhile, things are quiet as we move toward the lunch hour. It seems we should be done with this emergency in short order.

March 29, 1979: Afternoon and Evening

My observation at lunch is that things are relatively calm and most people are going about their business. The local media are doing a responsible job of informing without unnecessary hyperbole. Back in my office, I continue to work with my small staff on the planning and response process. Our initial review included identifying resources, the potential for additional problems, and how we could best respond to the situation if it degrades. At 1:00 p.m., we receive a telephone call from PEMA. The substance of the call was more of a rehash with no significant change. We immediately pass this latest information to the locals. At approximately 3:45 p.m., the teletype starts spewing the latest information from PEMA. It details where we stand up to 2:00 p.m. on March 29th. Basically, the message states: Met Ed advised at 10:00 a.m. the situation is secure, cooling in process, no danger to the public and stabilization of the reactor is expected sometime today. We double-checked the information via telephone call. PEMA advised the state EOC would remain operational and no further Sitrep would be sent unless conditions warrant. Jack Semanko, the head of the ambulatory portion of Harrisburg Hospital, which was a block away, came in for an updated briefing. After sharing what information I had, we agreed that it appeared nothing that could or would impact the hospital, or any of the other hospitals we had in Dauphin County, was imminent.

Later that afternoon, I received a call from Mike Pintek of WKBO radio. Mike told me he had a Dr. Sternglass in his studio. The doctor was recommending that pregnant women leave the area. I told Mike I was not well versed enough to comment, but would check with both the State Civil Defense and Bureau of Radiological Protection. Mike advised me that he had a deadline and was going to put Sternglass on the air soon. By the time someone from the state got back to contacting the station, Dr. Sternglass was already on the air. His comments just added to the tension. I did learn that Dr. Sternglass was from the Pittsburgh area and, depending on who you talked to, he may or may not have been an "expert" in this field. We soon received many calls from pregnant women. Obviously, questions dealt with the TMI situation and what danger were they in. We advised them to stay indoors, stay calm, and call their doctors. There were many other anti-nuclear groups and individuals who took advantage of the situation. To this day, I think there is a time and place for everything: Not now!

We at Civil Defense were very busy dealing with planning and normal day-to-day police, fire, and ambulance calls; we did not have time to engage in this speculation and biased agendas. It was not helpful; we were trying to keep people calm and up to date, while others were intent on just pushing the anti-nuclear message regardless of the impact it had on the population. Today, with the advent of social media and a lack of good emergency/public information programs, the negative impact of an emergency or disaster on the public can be troublesome.

Updating plans, checking resources, trying to anticipate problem areas, and keeping the local officials and first responders informed kept us busy through the remainder of the day. Several reporters from various media outlets stopped in and asked for updates. We told them things seemed calm. One asked me, "If things are so calm, why are you still in the office?" Some seemed convinced that I knew something. I did not. Well over 40 hours without sleep, I thought I could go home tomorrow. I thought by the end of Friday, March 30th, the accident should be over. Little did any of us know what type of day it would turn out to be and how it would ultimately change so many lives.

Friday, March 30, 1979

At 2:30 a.m., a call from the state indicates there apparently has been a discharge into the Susquehanna River. Nothing serious and I was asked to pass the information to the local agencies. I call for an update around 6:50 a.m. I am advised things are improving. Discharging stopped, monitoring is continuing and the reactor is not in cold shutdown yet. I pass updated information to local officials who are ready to respond to any contingency. About one hour later, the state advises me that some water with low levels of xenon is being pumped in a controlled manner into the river. I am also advised the reactor temperature is coming down. Once again, the information is given to the local officials. Several local directors are rather irate and tell me they are losing money not going to work and that a lot of responders are upset. I am also asked if there is no problem, why are we updating plans, still on standby, and why are you still in your office. Good questions with no good answer. 8:15 a.m. dispatchers change shifts and things seem normal. However, in the next two hours, a unique series of events will take place that will shake the entire state, the country, and the world.

8:35 a.m. I receive a call from Jim Floyd, an employee at the Three Mile Island site. Our conversation is very brief. He said, "We have a problem at the site. I need to get in touch with PEMA fast." I tell him I will take care of it. Floyd's voice conveyed a sense of urgency. I call PEMA, tell them to call Jim Floyd, and advise them that there may have been a release of more radiation into the environment. Concluding the telephone call, my staff and I immediately start to get word out to all of the local officials. I advise the commissioners we might be on the verge of major problems. I brief them on the current situation and the potential that there could be off-site consequences. Each commissioner asked several questions about the plant itself and where did "we" stand in terms of response. I remember Commissioner Minnich saying to me, "Kevin, it's your baby, we'll help in any way but will stay out of the way." The commissioners then returned to their respective offices.

Jim Cassidy from PEMA called at 8:54 a.m. He advised that as of 8:51 a.m., there was an on-site emergency declared at TMI. Significant increase in the readings to 1.2 Rem or 1200 mR/hour. He said no actions were necessary at this time, and to standby for additional information. Jim also advised me the 1.2 R/hour reading was taken at 600 feet above the stack and a reading of 14 mR/hour was recorded at the site boundary fence. We immediately relay this information to local elected officials, police, fire, and civil defense personnel. Internally, we went into extra high gear to review the exact actions we would take at the county level should we have to coordinate the evacuation of a five-mile area around the plant. I called Tom Blosser, who was the Cumberland County Civil Defense Director, and brought him up to date on where we stood. We discussed the type and level of assistance he could provide in the event of an evacuation. Tom assured me that I could count on whatever assistance I needed. The same was true of other counties offering assistance. At that time, this activity was the normal way of operating. This was to become one of the larger lessons learned. Mutual aid was a critical part of planning and response. Currently across the United States, we now have the Emergency Management Assistance Compact (EMAC). The compact has been signed by all 50 states and provides specific clarity and details on providing assistance to other states when an emergency/event has not reached the presidential declaration stage. Ironically, it would be a few hours later and Cumberland and four other counties would be developing their own plans for evacuations that might result from the TMI accident.

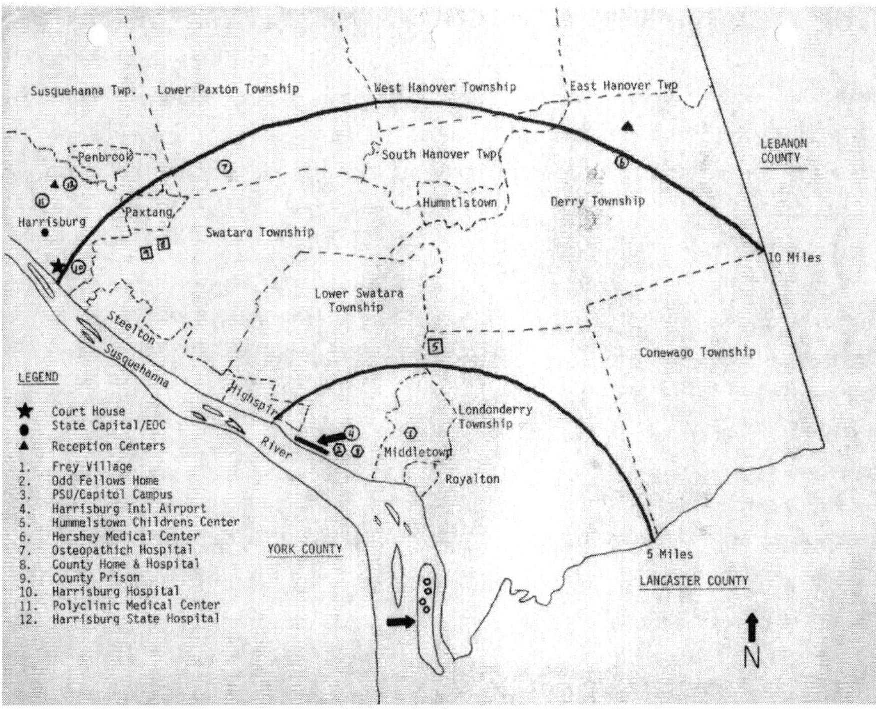

Figure 3.1. Three Mile Island Area Map.

Source: From Dauphin County emergency planning documents for Three Mile Island at the time of the 1979 accident. Courtesy of Kevin J. Molloy.

Note: Item no. 3 on the map is Penn State Harrisburg Campus, which today hosts the Academic and Administrative Home of The Pennsylvania State University's Intercollege Master of Professional Studies in Homeland Security Programs, where this book was edited.

At 9:40 a.m., I received a call from Ken Henderson, the PEMA Director. His voice and message conveyed an urgent and serious tone. He told me that the situation had deteriorated to the point that an evacuation was imminent. He went on to say that I should expect a call from the governor's office in a few short minutes to be told to evacuate a ten-mile distance as opposed to five miles. I was stunned momentarily. Not necessarily at the word evacuation, rather the words ten miles. I told him we would take care of five miles first and then work out to ten miles. I hung up and immediately began to issue orders. The commissioners were asked to come to the EOC, and I requested the Sheriff's Department to secure the entrance way to the EOC and Communications centers. John was dealing with our staff and I had the dispatchers alert all units over the police, fire, and ambulance frequencies. I called WHP Radio-TV, our Emergency Broadcast (EBS) station, and was immediately put on the air. I advised residents in communities near the plant that while not evacuating at this time, they should start to prepare to move if necessary. It was shortly after that broadcast that the telephone system crashed for a period of time. I took a good deal of heat for that decision. I would do it over again.

Putting this in perspective, in Dauphin County (see map in Figure 3.1), we went from a five-mile population of approximately 18,000, to a ten-mile population of some 127,000 residents, and finally to a 20-mile population of 250,000. The number of hospitals potentially affected by the TMI accident increased from zero to six. Over a half dozen nursing homes were in the 20-mile bull's-eye, as was the county prison and detention home. This is illustrated by the county statistics in Figure 3.2.

Over the next several days, problems multiplied in terms of numbers and complexity. This was extremely significant. The loss of local response resources was a major blow to our efforts. We now had to receive major support from across the state, which was manpower and time intensive. This new level of mutual aid and assistance is similar to the Emergency Management Assistance Compact that exists today.

COUNTY STATISTICS

	DAUPHIN	YORK	LANCASTER	CUMBERLAND	PERRY	LEBANON
POPULATION						
5 miles	28,000	10,380	7,000			
10 miles	127,000	38,460	20,000	30,818		
20 miles	200,339	235,665	110,000	88,096	9,000	37,000
HOSPITALS						
5 miles	TOTAL-6	TOTAL-3	TOTAL-1	TOTAL-2	TOTAL-0	TOTAL-0
10 miles	4					
20 miles	6	3	1	2		
NURSING HOMES	TOTAL-13	TOTAL-10	TOTAL-13	TOTAL 14	TOTAL-1	TOTAL-5
5 miles	2					
10 miles	5			4		
20 miles	6	10	13	10	1	5

School Dist
3
6
96

Figure 3.2. County Statistics of Population, Hospitals, and Nursing Homes.
Source: From Dauphin County emergency planning documents for Three Mile Island at the time of the 1979 accident. Courtesy of Kevin J. Molloy.

TMI as the "First" Emergency

The Three Mile Island accident is considered by many to be the "first" major emergency of modern times. There is a litany of firsts associated with the accident. Obviously, it was the first nuclear accident in a populated area. Other firsts include:

- It was the first time that the issue of moving multiple governments had to be addressed.
- It was the first time that local governments had to determine if they could perform local elected duties outside of the physical confines of where they were elected.
- It was the first time multiple hospitals and nursing homes would have to be relocated and the first time that some prisoners were released early as one method of dealing with the inmate population at a county prison.
- It was a first time necessitating the quick set up of an alternate emergency operations center and rumor control telephone lines.

While different types of events, the challenges faced during TMI, the September 11, 2001 terrorist attacks, hurricanes such as the devastating Katrina, as well as other disasters and emergencies, are quite similar. The emergency management community made significant changes based on lessons learned from each event in terms of planning, response, and recovery. In almost every after action report and case study of major disasters, a lack of communication, coordination, and lack of foresight are some common denominators. Yes, we improve and make changes, but we always seem to have these same underlying issues.

From an emergency management perspective, we recognize that certain events (i.e., terrorism, hazardous materials, etc.) will dictate specific actions be taken to resolve the situation. First responders and those involved in planning and management understand there is a basic response to every disaster, emergency event, or any pre-planned activity. The response to the 9/11 tragedy in New York City was for the most part similar to our approach to Three Mile Island. Size up the situation, know your resources, and deal with the totally unexpected. While the loss of life on 9/11 was more significant, the challenges facing emergency management and first responders were very much the same.

The advent of new technology and blossoming of social media have been beneficial. Major areas of concern remain to be constantly reviewed and addressed including inter agency communication and coordination issues, the ever-present and now immediate social and mainstream media. Establishing a common operating picture is more challenging than ever before. The ability to quickly and clearly define how "big is big and how bad is bad" is critical. It is something that is not focused on to the degree it deserves. To be successful, the emergency management community must clearly understand the need for a common operating picture to ensure critical coordination and utilization of resources. The problem of government coordination remains. This area is one of our more "open secrets" and something we do not care to acknowledge or address. It is a very real problem. We really should be talking, coordinating, and exchanging business cards prior to emergencies or disasters. Three Mile Island clearly demonstrated this lack of coordination between some segments of federal, state, and local government.

There is yet another "ticklish" subject. Perspective varies from where one is involved in dealing with Three Mile Island or any major disaster or event: The impact of politics, whether deliberate of inadvertent, is just bad. To this day, I respect Governor Thornburgh. However, he and I have much different perspectives on how he dealt with the TMI accident. Governor Thornburgh totally bypassed Emergency Management during the TMI event. He dealt with those with whom he felt more comfortable (political appointees, etc.). We (Emergency Management) as both a discipline and function remained in a constant state of limbo, making our job more difficult. I clearly remember State Director Henderson telling me, "Kevin, the governor's 'people' won't even talk to us." I was shocked. I made the County Commissioner aware of this situation and they were extremely upset. On Saturday, March 31st, State Senator George Gekas and State Representative Jeff Piccola came into my office and I advised them that I was going to take action if someone from the governor's office did not come down to the Dauphin County EOC. It was not until I literally threatened to go on the local media and tell those in Dauphin County to evacuate that we finally were recognized by the governor's office.

I was advised that legally I could not "order" an evacuation. I already was aware of that; however, I also knew that the residents of Dauphin County were going to listen to someone local, someone they knew, and not a bunch of newly elected government officials. Following that threat, Lieutenant Governor Scranton showed up at the EOC Sunday morning to discuss what was going on. We gave him a quick lesson in the reality of the situation from the local perspective. I think he was shocked and stunned by the magnitude of what we at the county level were dealing with and what we had actually accomplished. Evacuating the population, hospitals, nursing homes, and a prison was an eye-opener. Lieutenant Governor Scranton also learned how the unwillingness or inability of the state to coordinate with us at the boots-on-the-ground level was making our job more difficult and ultimately causing additional concern among the public. Following that meeting with the lieutenant, coordination improved to a limited degree.

Lessons Learned and Challenges ahead for Emergency Management in Homeland Security

I cannot emphasize the following statement enough: There is *no room* for politics when and where the safety of individuals is involved. I am not necessarily referring to party politics. Unfortunately, to this day, politics continue to come into play and negatively impact emergency management efforts both prior to and during emergencies and disasters. When disasters and/or emergencies reach a major magnitude status, the politics seem worse. Additionally, to this day, territorial issues remain in political (geographical) sub-divisions, cross disciplines (i.e., law enforcement, fire, and public works), and even within the same organizations themselves. This was sadly the case back in the days of Three Mile Island. Everyone wanted to be in charge. With the advent and increased use of the Incident Command System and following the direction of the National Incident Management System (NIMS), we have improved dramatically in eliminating the "who's in charge" issue. The problem still exists.

One other major issue that needs to be addressed is our loss of talent (corporate or industrial knowledge) in the emergency management field. This comes about for a variety of reasons, including lack of leadership, programmatic changes simply for the sake of change thus wasting valuable resources, and hiring practices. We must ensure that we are in fact bringing well-educated and experienced individuals into the emergency management field.

Hopefully, you better understand why my perspective remains that in the United States and internationally, we must revisit and constantly challenge how we do business in homeland security and emergency management. I believe the current perspective that we remain driven more by the latest situation or event that captures the attention of the media is very real. We need a more comprehensive approach. The "disaster du jour" mentality is alive and well in America and around the world. We must start looking at our programmatic responses to any event from a broader perspective. The main focus on planning and the expenditure of funds should be down at the local level. Monies spent on a large number of federal programs are, in fact, a drain on our limited resources.

Our use of technology as well as improved planning and coordination have helped save the lives of many individuals. Our use of technology has also created a vacuum in terms of continued reflection of past events, a continuance of the "We have always done it that way" and "It won't happen here!" philosophies. Any emergency plan or written procedure is a dynamic, fluid document. As such, they must be taken off the shelf frequently, and reviewed and revised as appropriate. Although plans and procedures may be officially approved, they simply are never set in concrete. It is incumbent on those who are entering the world of emergency management to understand not only the concepts, but take the time to put the books down and get one's hands dirty. The public that we serve demands and deserves our best.

Editor's Note

Penn State Harrisburg organized a conference to commemorate the 35th anniversary of the Three Mile Island accident in March 2014. The conference was entitled *"TMI@35—The Three Mile Island Nuclear Crisis in Perspective: Analyses, Stories, Policies,"* and proceedings are available online on http://sites.psu.edu/tmi35. They include, among other things, the conference brochure with suggestions for further reading, a conference report, as well as presentation slides and videos from keynote speeches and panels, also with contributions from Kevin J. Molloy.

4

Beyond the Storms: Implementing Smart Resilience in Turbulent Times

DANE S. EGLI

Introduction

There is no safe harbor to avoid the impact of catastrophic events—"physical" disasters, such as extreme weather, earthquakes, and terrorism, or those that are "virtual," such as chemicals, infectious diseases, money, and electrons—and their direct or indirect consequences on the homeland. We face rapidly changing times, globally and nationally, marked by complexities and uncertainties that force us to make difficult decisions about homeland security and community preparedness.[1] This chapter suggests a new approach that leverages whole-of-nation and private-sector capabilities to systematically strengthen preparedness and resilience.

The destabilizing impact of recent disasters has provoked a reactionary posture that is not necessarily in the long-term interests of community preparedness. Beyond the unquantifiable human costs associated with hazards, figures from 2011 reveal that economic damages from natural disasters in the United States exceeded $55 billion.[2] In 2012, there were many domestic disasters, including Superstorm Sandy, lethal wildfires, regional droughts, domestic terrorism,[3] and the spread of West Nile virus. While we must address immediate crises and apply the lessons learned from 9/11, Hurricane Katrina, the Haiti earthquake, and the Deepwater Horizon oil spill, this study looks "Beyond the Storms" of pressing events to identify strategic opportunities that would make the nation better prepared and more secure through a new focus on systemic resilience based on quantifiable metrics and standards.

This chapter documents the results of an exploratory study designed to examine the state of preparedness in the United States and its communities by drawing upon policy reviews, case studies, and expert interviews. It addresses the current state of preparedness based on policy directives; the complexity and interdependencies of critical infrastructures and central role of global supply chains; resilience as a public good enabled by collective action, interagency coordination, and public-private partnerships (PPPs); policy implementation under the current whole-of-nation interagency constructs; and ways to strengthen preparedness and resilience.

More than Critical Infrastructure Protection

The initial focus of the study was on Critical Infrastructure Protection (CIP), but quickly extended to the broader concepts of preparedness and resilience, as the research pointed toward a different strategic view. It is clear that disaster management planners and leaders across public-private sectors understand how to build and protect infrastructures within their areas of responsibility, but often lack awareness of the challenges facing adjacent geographic or infrastructure sectors. Therefore, there is a significant lack of resilience thinking, integrated planning, and collective action at the national, regional, state, and local level.

Empirical data revealed a fragmented planning and risk assessment process where agencies tend to focus narrowly on their areas of concern (ports, bridges, railways, tunnels, telecommunications, power, water, etc.), yet fail to recognize the complex interdependencies of the broader homeland security enterprise. Furthermore, when preparing budget requests or proposals for grant funding, submissions reflected this isolated approach that safeguarded their specific region or municipality at the expense of the collective good. Initial findings uncovered that while we live in an interconnected world of complex 21st century vulnerabilities, we often employ 20th century ways of thinking and planning for security—allocating resources for physical protection based on the most senior member on the congressional committee, best-written grant proposal, or a bias for "fixing things" to relieve the most immediate pain.

Current State of Preparedness

In the past century, catastrophic disasters have brought dramatic reminders that we need better understanding of risk mitigation and how to improve critical infrastructure resilience. It is important to recognize that the majority of all critical infrastructures,[4] across 16 interdependent sectors, are owned and operated locally or regionally by the private sector and often influenced by an uncoordinated mixture of investors, volunteers, non-profit organizations, labor unions, and utilities; and the federal government is often hard-pressed—or simply unable—to institute changes.[5]

It is increasingly clear that the nation needs a new resilience-based approach to critical infrastructure protection that can be integrated into the homeland security enterprise and builds greater trust across public and private stakeholders. Action plans are needed—not just government strategy documents—to implement preparedness policies and incentivize the private sector to support policymaking, equity investments, cost recovery, new insurance markets, and legislative changes, or the nation's infrastructure will remain in the current state of disrepair.[6] Researchers and global innovators emphasize the futility of popular security habits (with a fixation on physical protection) and assert the need for new, more pragmatic resilient approaches because the status quo is not working and change will not occur without a new emphasis on smart resilience.[7]

There is a growing demand among local and regional planners for a shift in disaster preparedness that places a practical emphasis on resilience. The statistics predict a looming challenge that is not well understood by a population that expects critical infrastructures automatically to serve their needs. For example, approximately 75% of the nation's GDP is fueled by our metro communities and over 80% of the U.S. population is located in 11 mega-regions, where new approaches to resilience thinking are needed most. The risk assessments and analytical frameworks that model the infrastructure resilience in an interconnected environment must be developed to better understand these complex urban communities, as underscored by the cascading regional impacts caused by disasters such as Superstorm Sandy of October 2012. Given that federal funds are receding, there is a need for increased private sector participation and regional approaches to address the "wicked problems" being faced by preparedness planners. The 90,000 local, regional, and state flood maps maintained by the Federal Emergency Management Agency (FEMA) are incomplete or outdated, as extreme weather conditions have revealed in recent years;[8] and corresponding products—risk maps—for the regional all-hazards threat environment do not exist.[9]

A Nation at Risk

The United States is at a critical juncture where many infrastructures are aging or failing, and current efforts to address the challenge are insufficient. America invests barely 2% of its GDP in infrastructure renewal and maintenance, one of the lowest figures among the world's industrialized nations. The following facts expose some of our demographic and infrastructure fault lines:[10]

- Using input from engineers in every state, the American Society of Civil Engineers (ASCE) gave the national infrastructure an overall grade of D-plus, a score impacted by failing areas such as aviation, drinking-water supply, roads, transit, and sewage treatment.
- 42% of the nation's urban highways are congested, which costs our economy more than $101 billion annually in lost time and fuel, while increasing harmful exhaust and pollution.
- 38% of the roads are in disrepair and the U.S. must invest $225 billion per year over the next 50 years to maintain and enhance a decaying surface transportation system; we are currently spending less than 40% of this amount.
- The U.S. operates 361 major ports and delivers 13 billion containers globally per year, supporting the world's largest trading commerce ($645 billion), yet there is no executive agent assigned to manage or integrate this dynamic global supply chain and associated policies.
- The Army Corps of Engineers estimates that over 95% of overseas trade produced or consumed by the U.S. moves through our nation's commercial ports. Yet, our ports are in danger of losing their competitive advantage due to the slow and very complex process of critical dredging projects.
- By 2020, every major U.S. container port is projected to be handling at least double the volume it was designed to handle, and by 2040, this figure is projected to triple.
- Over the next 20 years across the nation, railroads are projected to need nearly $200 billion in investments to accommodate increased freight traffic.
- Over 25% of the nation's bridges are considered structurally deficient.
- There are approximately 85,000 dams in the country, averaging over 50 years old, and 4,000 are considered structurally at risk.
- There are 500 major urban public transit operators, 66,000 chemical plants, over 2,800 electric plants, 104 nuclear power plants, 1,600 municipal waste water facilities, over 2 million miles of pipelines, 87,000 food processing plants, and 5,000 public airports owned and operated independently; however, they are highly interdependent in their functional capabilities.

The Issue of Our Day

This sampling of facts bespeaks a menacing challenge that must be addressed with a sense of urgency or infrastructure systems will continue to erode along with public safety and security. Many believe that this challenge—to dramatically transform how we secure and maintain our critical infrastructure through a posture of resilience—is the issue of our day. As the "Greatest Generation" confronted the threats from Germany and Japan in World War II, and proceeded to build the domestic infrastructure system that has been enjoyed for the past 60 years, the current generation also faces defining challenges. There is a need for a comparable resolve in facing today's serious challenges—including economic austerity, natural disasters, and global security threats—to forge a culture of preparedness and invest wisely in critical infrastructure resilience.

Globalization and expanded markets have unwittingly introduced the potential for greater cascading hazards. Many of the advancements and freedoms that accompany technology and social networking advancements have in fact made us more vulnerable. The lessons of globalization and cybersecurity suggest that vulnerabilities from "system complexity" have expanded at a faster pace than our means of mitigating them. Further, interconnected infrastructure systems, while offering greater speed and efficiency, now present larger targets for potential exploitation by criminals and terrorists because these networked architectures are operated in an open society.[11] The qualities of speed, access, and anonymity that make these robust systems so effective are the same features that make them more vulnerable to nefarious elements.

Threats to our physical and virtual security are complex, new, and growing. They demand a new approach to how we prepare for attacks and sustain our infrastructure, as well as how we mitigate, respond to, and recover from crises. Many of the threats we face *cannot* be deterred or countered simply because the provision of civil liberties and a free society are paramount. The idea that complete physical security is normal, or even achievable with enough Transportation Security Administration (TSA) agents and Customs and Border Protection (CBP) officers, is an illusion that is consuming our finite economic resources.[12] We must direct our efforts to building a new vision of 21st century resilience in the face of uncertain and complex risks.[13]

Hope and Hubris

This new vision starts with a careful examination of cultural mindsets and deeply held attitudes toward emergency hazards. All disasters are personal, and the first priority in preparedness planning is an *individual* responsibility. According to FEMA guidelines, people should be self-sufficient for 72 hours after a disaster strikes if evacuating, and longer if sheltering-in-place.[14] That is where the breakdown often begins. Experts emphasize that America will never truly mitigate the harsh impact of disasters until the public takes personal ownership for the preparations necessary to respond to and recover from emergencies.[15]

The challenge of raising public awareness before a disaster strikes is manifested in an attitude of "It will not happen to me," or "We will just deal with it at the time." While clearly a dangerous reaction, as learned from hurricanes Katrina, Irene, and Sandy, it can also be linked to difficult economic times. A society struggling with poverty and unemployment is focused on financial survival and is, therefore, less likely to spend essential money (or even discretionary funds) on insurance or redundant systems that do not have an immediate return on investment (ROI) or help to pay the bills. Yet, research indicates there is a 4:1 ratio associated with preventive action supporting infrastructure; for every dollar spent now on resilience-building and disaster preparedness, one can avoid at least four dollars in future losses. However, an investment in resilience is often considered an "invisible" payoff because if built and maintained properly, infrastructure works as planned and life continues uninterrupted.[16]

In addition to an attitude of denial, many people simply refuse to take ownership for individual preparation because "It is the government's responsibility, and FEMA will take care of us." This attitude must be challenged through a combination of initiatives at the local, regional, state, and national levels.[17] Because extreme weather, terrorist attacks, and health epidemics are inevitable, citizens should rely less upon outside assistance and depend more upon their personal resources, enabling communities to be better prepared for future disasters. Additionally, rather than creating a new program that would add another financial burden to already-struggling families and small-business communities, there should be incentives that can be applied to existing (family and small business) financial budgets to reward action taken to support resilience. Although federal funds are needed to implement some programs, many requirements can, and should, be accomplished through state-level programs.[18]

Moving beyond human factors to an all-domain context (land, maritime, air, space, and cyber), the study proposed a framework that draws upon the current pillars of national preparedness (prevent, protect, mitigate, respond, and recover), offering a continuum of options from infrastructure protection to functional continuity under a broad umbrella of resilience (see Figure 4.1).

Figure 4.1. Resilience Framework.

This framework suggests a view of resilience that helps define the planning variables involved in national preparedness. There is an important distinction between "protection," which focuses on the *threats*, and "continuity," which focuses on *resuming operations* despite the consequences. In this framework, both objectives are enabled through the operating principles of *resilience*. Moreover, while prevention and protection may be the preferred options in many cases, they are not always possible. Therefore, an increased focus is needed on mitigation, response, and recovery.[19]

Understanding Resilience

Resilience, in a physical and structural sense, is considered the ability to bounce or spring back into shape or position after being pressed or stretched. However, the broader concept of resilience originated in the ecological and social sciences, where it is critical for survival and growth within complex systems. Research on these systems suggests that they perpetually evolve through an "adaptive cycle" of growth, crisis, transformation, and renewal. Resilience is not only the ability to recover from disasters and flex instead of snapping, but is also the ability to *get stronger* as a result of adversity.[20]

The traditional view of physical security and infrastructure protection involves preparing for risks and dangers we do not know about, hardening facilities against a potential attack, and adding more redundancies and defensive layers; analogous to an individual saving money in case of job loss. However, resilience suggests a different type of preparedness where one would, in addition to saving money, learn new skills and establish a broader network to land a better job.[21] In the case of critical infrastructure, rather than "fixing things" or adding more safeguards through the congressional appropriations and authorization process, we should systematically evaluate where and how we can make optimal investments to rebound from a disaster, and in some cases, pre-position recovery resources and have trained response teams on stand-by.

Among the various perspectives on resilience held by engineers and social scientists, the Multidisciplinary Center for Earthquake Engineering Research (MCEER) offers common actions to achieve resilience across all infrastructure sectors that are arguably inadequate for modern levels of complexity: robustness, redundancy, resourcefulness, and rapidity. These traditional measures signal the need for innovative thinking that goes beyond simply investing in more "guns, gates, guards, and locks" in order to protect physical structures. We must identify the capabilities needed to mitigate the impact of, and respond to, *inevitable* hazards. Oftentimes, it is nearly impossible to determine the risk of hazards because there is too much uncertainty to quantify the threats, vulnerabilities, or consequences. The variables in the equation can be so complex that this information is sometimes unknowable.[22]

Therefore, the best return on investment—and source of confidence—across interdependent supply chains, infrastructure sectors, and interconnected systems is not a fortress-protection mentality, but an investment in "functional resilience" that instills a level of confident anticipation and personal preparation for inevitable disasters looming over the horizon. In a post-9/11, post-Sandy, post-H1N1 ("swine influenza") environment, we should no longer be surprised by disastrous interruptions to our otherwise normal lives, but should anticipate, and even expect, them. Our communities should prepare for disruptive events with a resilient mindset—even if it does not come naturally—because there is a value proposition to be realized.

Economics of Resilience

The degree of interdependence across critical infrastructure sectors has been amplified by globalization, advanced technologies, and supply chain pressures. Our team at Johns Hopkins University Applied Physics Laboratory is studying—through modeling, analyses, and empirical research in places such as the Port of Baltimore and Austin, Texas—the measurable impact of disruptive events, governance, and societal demands upon resilience ecosystems in bounded geographic areas.

Governments, communities, and individuals are not helpless in the face of natural disasters like Typhoon Haiyan, the category five super typhoon that struck the Philippines in November 2013, killing thousands and displacing hundreds of thousands. There are practical safeguards that can be designed within the multidisciplinary worlds of engineering, cyber-physical, and the social, behavioral, and economic (SBE) sciences if we systematically identify the independent variables that contribute to critical infrastructure interdependencies, conduct analyses that support a generalizable model, and test these methods under simulated and real-world conditions. Drawing from the principles of collective action theory and computational analytics our studies are seeking to quantify the cost accounting and value proposition behind resilience by integrating economic factors into the research.

By creating a more connected world, globalization and technology have increased transparency and business efficiencies while simultaneously making systems more vulnerable, amplifying the consequences of small local disruptive events as well as high-impact but low-probability events, referred to as black swans. A groundbreaking 2005 study by Kevin Hendricks and Vinod Singhal analyzed the effects of 827 disruption events. The study found that over the course of three years, the average disruption reduced stock returns by up to an incredible 40%. The result was a negative regardless of a disaster's cause.[23] A follow-up study showed that disruptions increase share price volatility by 13.5%, reduce operating income by 107%, decrease sales growth by 7%, and increase costs by 11%.[24] Infrequent and unlikely disruptions thus can, in a moment, destroy value created over a long period. As the study asserted, "There is a direct relationship between efficiency and risk."[25]

Supply-chain disruptions of varying degrees of severity are common. Of the respondents of the Business Continuity Institute's 2012 Annual Supply Chain Resilience Survey, 73% experienced at least one supply-chain disruption. Of these, nearly 40% occurred below the immediate tier-one supplier, showing the interconnectedness and complexity of modern business practices. Interestingly, information technology and telecommunications outages were the top sources of disruption, with severe weather taking a close second. The primary consequences of these disruptions are loss of productivity, increased cost of work, loss of revenue, and customer complaints.[26]

Therefore, globalization and supply-chain efficiencies, while among the great advances of the modern era, are only part of the value equation. Just as important is supply-chain *resilience*: the ability to withstand a crisis, absorb damage, recover quickly, and adapt to disruptive events. Resilience requires long-term planning and investment in redundancy, interoperability, and agility. Disruptions often cannot be predicted or controlled, but their negative impacts are incontrovertible. As Hendricks and Singhal conclude, "Investments in increasing reliability and responsiveness of supply chains could be viewed as buying insurance against the economic loss from disruptions."[27] In addition to mitigating the risks and hazards of supply chain disruptions, resilience helps prepare businesses for future market slumps. According to Morgan Swink of the Neeley School of Business, "A firm's ability to weather economic downturns, deal with volatility, and manage costs under shrinking demands depends in large part on the resiliency of its supply chains."[28]

Resilience is "disaster agnostic," meaning it will favorably mitigate all forms of damage, to varying degrees, caused by earthquakes, terrorists, pandemics, and economic downturns. Though it may be difficult to quantify, after every disaster businesses that prepared ahead of time come out on top.[29] For example, an earthquake three years before the 2011 Japanese Fukushima tsunami helped prepare a semiconductor manufacturer to recover before its competitors because it had established a strategy to shift production to unaffected manufacturing plants.[30] Maintaining critical operations in the face of disruptive events confers a measurable competitive advantage in the marketplace.

We know from the emerging national policies and governance that investing in resilience is a national imperative and increasingly considered a basic business practice. In addition to mitigating disaster-related damage by introducing new flexibility, it increases productivity, revenue, reputation, and shareholder value.[31] Investing in resilience before disaster strikes is the smart choice for individuals, companies, and governments alike. What is the value proposition or return on investment? For individuals, it is an investment in adaptive safety and security. On the part of the government, it saves lives and property. For businesses, it protects the bottom line and sharpens their competitive advantage.[32]

Many Creative Ideas and Partners Available

The encouraging news is that we are not alone. There are international partners who face similar challenges and are willing to share information, and professional fora to draw upon academia, public-private partnerships, and the experience of commercial industry to improve preparedness. We can no longer afford—financially, technologically, or environmentally—to take a unilateral approach toward preparedness when facing a globalized system of interdependent supply chains and infrastructure sectors.[33] Below are some exemplars that may help inform the efforts of preparedness planners in the public and private sectors:

- Military and interagency studies on unrestricted, irregular, and unconventional conflict provide a relevant body of literature and critical thinking because of the similarities to complex and asymmetric factors that are present within critical infrastructure resilience and all-hazards environments.
- Academic institutions and Homeland Security Centers of Excellence across the country are actively studying and testing many of the issues surrounding preparedness and critical infrastructure resilience, generating research-based models to help inform strategic and operational planning.
- After the attacks of 9/11, aviation transportation was grounded for three full days as a security precaution; and the maritime ports, with the exception of the Ports of Los Angeles and Long Beach (POLALB), California, were closed, disrupting vital commerce and trade. Actions by the POLALB provide an example for centers of intermodal activity and complex interdependencies during future incidents, where advanced planning communications among emergency management, law enforcement, port authority, regulators, labor unions, and congressional leaders allowed increased security and sustained port operations to occur simultaneously.
- Counterparts to U.S. homeland security planners in the British Cabinet Office's Civil Contingencies Secretariat have successfully implemented "business continuity" strategies with the private sector by leveraging the relationships that small and medium-size enterprises already trust and rely upon: suppliers, customers, banks, insurance companies, professional organizations, and distributors.
- The Australian National Security Strategy reflects a *Beyond the Storms* approach, placing "ensure a safe and resilient population" and "secure our assets, infrastructure, and institutions" as two of their four strategic national security objectives. One of the eight pillars to support these

objectives is "strengthening resilience" across their society. Recognizing "it is not possible to eliminate all risks to national security," they place a focus on public-private partnerships and building community cohesion in order to "respond and recover quickly" to restore essential services.[34]

The Implementation Challenge: Public-Private Partnerships

The key to implementation of any local, state, regional, or national level policies in support of critical infrastructure resilience, preparedness, or business continuity is leveraging of public-private partnerships. Most public policy is generated by federal or state level agencies, yet most disaster management and emergency response activity occurs at the local level among private sector owners and operators. Since the majority of critical infrastructures are managed by the private sector, the greatest potential advancements in community resilience are often found among private-sector stakeholders.

Strategic policies and national strategies assert the importance of expanding public-private partnerships (PPPs) and acknowledge the need to incentivize venture capitalists in supporting infrastructure improvements and disaster preparedness; but few details have emerged in academia, think-tanks, or public policies identifying *how* to incentivize these communities, infuse private sector investments, or significantly expand PPPs. Private industry is in the business of making money by seeking investments that will reduce risk, demand less manpower, and provide a reliable flow of revenue in the current fiscal climate. As the public sector seeks to remove barriers and incentivize the private sector to increase participation in joint ventures and partnerships, there will be a natural resistance to form PPPs within commercial industry because anything that could provide competitors with a market advantage is not something they will be interested in.

National Call to Action

Drawing from study findings, a team of researchers, analysts, and planners is taking steps to address some of the major challenges outlined in this chapter by implementing a three-phase Resilience Implementation Process (RIP). Hopefully, these efforts will serve as a catalyst among public and private sector leaders to operationalize critical infrastructure resilience. The objective is to develop a systematic methodology that can be generalized or adapted to other communities in a flexible and scalable manner as reflected in the framework illustrated in Figure 4.2 below.[35]

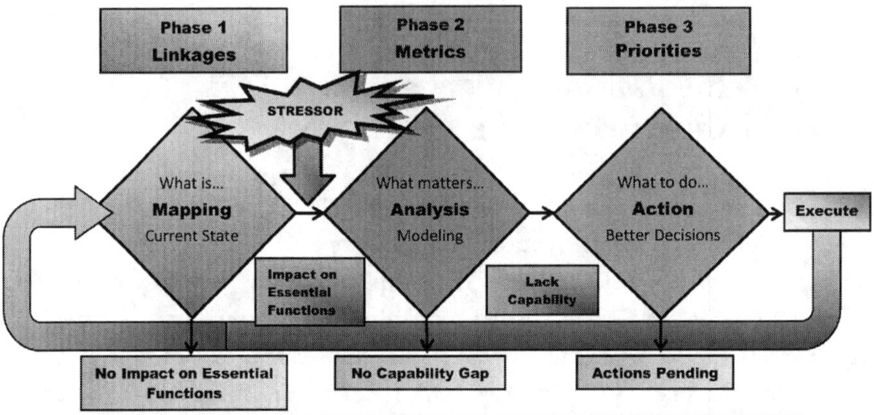

Connectedness	Measurement	Operationalize
Physical & Virtual	Qualitative & Quantitative	Materiel & Non-materiel
Dependencies	Criticality	Governance, Operations, Systems
Interdependencies	Capabilities	Time, Cost, Scope

Figure 4.2. Resilience Implementation Process (RIP) Framework.

The three phases of the Resilience Implementation Process Framework are mapping, analysis, and action, as described in the following:

- The *first phase* is systematic *mapping* of infrastructure dependencies and interdependencies in specific geographic areas to understand *what* are the essential linkages and functions for selected infrastructure sectors. This mapping is accomplished by collecting data from sources in the area or sector being studied. The outcome of this effort is a model that shows the relationship across sectors, intermodal variables, and supporting interdependencies.
- The *second phase* requires *analysis* to quantify options, using an integrated approach, which helps planners identify *how* to proceed, and prioritize capability gaps in preparedness and resilience. Measurements—derived from qualitative and quantitative assessment criteria—are obtained from testing the model. These findings will serve as key input to implement

actionable items for that geographic location or infrastructure sector. By applying a disruptive event or stressor, one can understand the potential impact of a disaster and begin to visualize connectedness between critical infrastructures;

- The *third and final phase* yields *action* plans to operationalize the outcomes from the first two phases, providing planners with leading indicators *where and when* events must be executed. During this phase, policymakers must forge a consensus among private- and public-sector leaders regarding discrete projects, priorities, and implementation efforts. This step requires feedback during the earlier phases and enables better decisions and priorities to achieve resilience through best practices and mitigating actions in a resource-scarce environment; and may require new action in governance, operations, or technology systems.

The RIP offers a general framework to inform future research of risk mitigation and critical infrastructure resilience. It suggests a practical approach that can be adapted to various geographic regions and infrastructure sectors through measurement, computational analytics, and modeling. These efforts will require extensive prototyping and testing, but are crucial to demonstrating the measurable impact of homeland security policies and the ability to operationalize preparedness and resilience within our communities.

Conclusion

This study highlights the need for action in promoting resilience at local, state, regional, and national levels, which can be accomplished through awareness, planning, resource allocation, and strong leadership.[36] Further, a systematic mapping of local, regional, state, and national capabilities-based requirements, based on a functional decomposition of all infrastructure dependencies and interdependencies, is needed. The findings of this study align with other major policies and initiatives, including Presidential Policy Directive 8 (PPD-8),[37] Presidential Policy Directive 21 (PPD-21),[38] the report "Improved Disaster Resilience Is Imperative to U.S.,"[39] and the STRONG Act legislation.[40] These policies represent a unique opportunity to build a resiliency consensus among public and private-sector planners.

While these documents contain many strong recommendations, they *will not make a difference* when the next disaster impacts critical infrastructure systems unless leaders agree to collectively join with cross-government interagency, private sector, and academic actors to form a coalition of like-minded innovators. Disaster management and emergency planners must be willing to move "Beyond the Storms" of the immediate budget crisis and pressure to simply react to the most recent disaster, and team with others to resource, implement, and measure this homeland security imperative. It will require strong leadership to bridge the current divide between individual rhetoric and collective action to invest in resilience.

Notes

The content of this chapter is based upon research conducted for the following book: Dane S. Egli, *Beyond the Storms: Strengthening Homeland Security & Disaster Management to Achieve Resilience* (Armonk, NY: M.E. Sharpe, 2014).

1. The President of the United States, *National Security Strategy* (Washington, D.C.: The White House, February 2015), https://www.whitehouse.gov/sites/default/files/docs/2015_national_security_strategy.pdf; Thad Allen, "The Future of Homeland Security: The Evolution of the Homeland Security Department's Roles and Missions," Testimony before the U.S. Senate Committee on Homeland Security and Government Affairs, July 22, 2012.
2. National Academies, Committee on Increasing National Resilience to Hazards and Disasters, Committee on Science, Engineering, and Public Policy, *Disaster Resilience: A National Imperative* (Washington, D.C.: The National Academies Press, 2012).
3. One NGO estimates there were 1,274 anti-government groups at the end of 2011 that can be regarded as terrorist, up from 131 groups four years ago.
4. "Critical infrastructures" are systems and assets, whether physical or virtual, so vital to the nation that their incapacity or destruction would have a debilitating impact on national security, economic security, public health or safety, or any combination of those areas; see Uniting and Strengthening America by Providing Appropriate Tools Required to Intercept and Obstruct Terrorism (USA PATRIOT) Act of 2001, Section 1016e, http://www.gpo.gov/fdsys/pkg/BILLS-107hr3162enr/pdf/BILLS-107hr3162enr.pdf
5. Stephen Flynn, *The Edge of Disaster: Rebuilding a Resilient Nation* (New York: Random House, 2007); National Infrastructure Advisory Council, *Cross Sector Interdependencies and Risk Assessment Guidance*, January 13, 2004, http://www.dhs.gov/xlibrary/assets/irawgreport.pdf
6. Allen, "The Future of Homeland Security."
7. Andrew Zolli and Ann Marie Healy, *Resilience: Why Things Bounce Back* (New York: Free Press, 2012).

8. U.S. Department of Homeland Security, Office of the Inspector General, "Challenges in FEMA's Flood Map Modernization Program," report no. OIG-05-44 (September 2005).
9. U.S. Government Accountability Office, *Critical Infrastructure Protection: Update to National Infrastructure Protection Plan Includes Increased Emphasis on Risk Management and Resilience* (Report to Members and Committees of Congress, GAO-10-296, March, 2010); John Morton and Catherine Feinman, *DomPrep Action Plan: Building Resilient Regions for a Secure and Resilient Nation* (Severna Park, MD: IMR Group, Inc., 2012); National Infrastructure Advisory Council, *Regional Resilience Scoping Study*, March 23, 2012, http://www.dhs.gov/xlibrary/assets/niac/niac-regional-resilience-white-paper-032412.pdf
10. U.S. Government Accountability Office, *Critical Infrastructure: Challenges Remain in Key Sectors* (Testimony before the Subcommittee on Homeland Security, Committee on Appropriations, House of Representatives, GAO-07-626T, March, 2007); Mark Gerencser et al., *Megacommunities: How Leaders of Government, Business, and Non-Profits Can Tackle Today's Global Challenges Together* (New York: Palgrave Macmillan, 2008); Daniel Byram, ed., *Introduction to Homeland Security* (New York: Pearson Custom Publishing, 2009), 164–165; Allen, "The Future of Homeland Security;" U.S. Department of Treasury, "A New Economic Analysis of Infrastructure Investment: A Report Prepared by the Department of Treasury with the Council of Economic Advisors," March 23, 2012, http://www.treasury.gov/resource-center/economic-policy/Documents/20120323InfrastructureReport.pdf
11. David Mussington, "Concepts for Enhancing Critical Infrastructure Protection: Relating Y2K to CIP Research and Development," *Monograph Report*, RAND Corporation Science & Technology, 2002.
12. The 9/11 hijackers spent less than $1 million to attack the United States; while the cost to try to prevent similar future attacks (in police, airport security, and monitoring systems) costs Americans a million dollars *an hour*.
13. Stephen Flynn and Sean Burke, *Critical Transportation Infrastructure and Societal Resilience* (Washington, D.C.: Center for National Policy, 2012).
14. "New App from Insurance Information Institute Helps Users Prepare for a Catastrophe before Disaster Strikes," Insurance Information Institute, published August 14, 2012, http://www.iii.org/press-release/new-app-from-insurance-information-institute-helps-users-prepare-for-a-catastrophe-before-disaster-strikes-081412
15. U.S. Department of Homeland Security, *National Response Framework*, 2008, http://www.dhs.gov/xlibrary/assets/NRF.pdf; U.S. Department of Homeland Security, *National Infrastructure Protection Plan*, 2009, http://www.dhs.gov/xlibrary/assets/NIPP_Plan.pdf; Federal Emergency Management Agency, *A Whole Community Approach to Emergency Management: Principles, Themes, and Pathways for Action*, FDOC-104-008-1 (Washington, D.C., December 1, 2011), http://www.

fema.gov/media-library-data/20130726-1813-25045-0649/whole_community_dec2011__2_.pdf

16. American Society of Civil Engineers, *Guiding Principles for the Nation's Critical Infrastructure* (Reston, VA: ASCE Publications, 2009); John Kerry, Frank Lautenberg, and Kirsten Gillibrand, "Strengthening the Resiliency of Our Nation on the Ground (STRONG) Act (S-3691)," introduced December 19, 2012, http://beta.congress.gov/bill/112th-congress/senate-bill/3691
17. Donald Kettl, *System under Stress: Homeland Security and American Politics* (Washington, D.C.: CQ Press, 2004).
18. U.S. Government Accountability Office, *Critical Infrastructure Protection: Update to National Infrastructure Protection Plan Includes Increased Emphasis on Risk Management and Resilience* (Report to Members and Committees of Congress, GAO-10-296, March 2010); "Web Users Increasingly Rely on Social Media to Seek Help in a Disaster," American Red Cross, published August 9, 2010, http://www.redcross.org/portal/site/en/menuitem.94aae335470e233f6cf911df43181aa0/?vgnextoid=6bb5a96d0a94a210VgnVCM10000089f0870aRCRD; U.S. Department of Homeland Security (DHS), *National Response Framework*, 2008, http://www.dhs.gov/xlibrary/assets/NRF.pdf; John Morton and Catherine Feinman, *DomPrep Action Plan: Building Resilient Regions for a Secure and Resilient Nation* (Severna Park, MD: IMR Group, Inc., 2012).
19. John Moteff, *Critical Infrastructure Resilience: The Evolution of Policy and Programs and Issues for Congress* (Washington, D.C.: Congressional Research Service, August 23, 2012).
20. Lance H. Gunderson and Crawford S. Holling, eds., *Panarchy: Understanding Transformations in Human and Natural Systems* (Washington, D.C.: Island Press, 2002).
21. Joshua Cooper-Ramo, *The Age of the Unthinkable: Why the New World Disorder Constantly Surprises Us* (New York, NY: Back Bay Books, 2009).
22. M. Bruneau, et al., "A Framework to Quantitatively Assess and Enhance the Seismic Resilience of Communications," *Earthquakes Spectra* 19 (2003); Cooper-Ramo, *The Age of the Unthinkable*; Zolli and Healy, *Resilience.*
23. Kevin B. Hendricks and Vinod Singhal, "An Empirical Analysis of the Effect of Supply Chain Disruptions on Long-Run Stock Price Performance and Equity Risk of the Firm," *Production and Operations Management* 4, no. 1 (2005), http://www.wilfridlaurieruniversity.ca/documents/17398/scm_glitches_and_shareholder_risk.pdf
24. Kevin B. Hendricks and Vinod Singhal, "The Effect of Supply Chain Disruptions on Long-Term Shareholder Value, Profitability, and Share Price Volatility," *Supply Chain Magazine* (June 2005), http://www.supplychainmagazine.fr/TOUTE-INFO/ETUDES/singhal-scm-report.pdf. A 2013 study by Accenture found that supply-chain disruptions reduce the share price of affected companies by 7% on average, see: World Economic Forum, *Building Resilience in Supply Chains* (January 2013),

http://www3.weforum.org/docs/WEF_RRN_MO_BuildingResilienceSupplyChains_Report_2013.pdf

25. Hendricks and Singhal, "The Effect of Supply Chain Disruptions," 3.
26. Business Continuity Institute, *4th Annual Survey: Supply Chain Resilience 2012* (November 2012), http://www.zurich.com/internet/main/sitecollectiondocuments/reports/supply-chain-resilience2012.pdf
27. Hendricks and Singhal, "An Empirical Analysis," 51.
28. "Weathering the Storm: How to Achieve Strategic Resilience through Supply Chain Excellence," Texas Christian University, Neeley School of Business, published June 15, 2012, http://www.neeley.tcu.edu/News_and_Events/Press_Releases/Weathering_the_Storm__How_to_Achieve_Strategic_Resilience_Through_Supply_Chain_Excellence.aspx
29. Mukta Agrawal and Casey Church, "Resilience Return on Investment—An Impossible Argument?" *Analytic Service, Inc.*, April 24, 2012, http://tisp.org/index.cfm?pk=download&pid=10261&id=12606%E2%80%8E
30. Kelly Marchese, Siva Paramasivam, and Michael Held, "Bouncing Back: Supply Chain Risk Management Lessons from Post-Tsunami Japan," *Industry Week*, March 9, 2012, http://www.industryweek.com/global-economy/bouncing-back-supply-chain-risk-management-lessons-post-tsunami-japan
31. "Business Case for Supply Chain Security and Resilience," U.S. Resilience Project, March 1, 2012, http://www.usresilienceproject.org/workshop/participants/pdfs/USRP_Resources_Chapter_6_030112.pdf
32. This section, "Economics of Resilience," is adapted from an article co-authored by Dane [S.] Egli and Jared McKinney, published in Center for Infrastructure Protection and Homeland Security, *The CIP Report* (Arlington, VA, George Mason University, November 2014), http://cip.gmu.edu/wp-content/uploads/2013/06/December2014_CIPHS_TheCIPReport_InvestmantsandAssessments.pdf
33. National Infrastructure Advisory Council, *Best Practices for Government to Enhance the Security of National Critical Infrastructures,* April 13, 2004, http://www.dhs.gov/xlibrary/assets/niac/NIAC_BestPracticesSecurityInfrastructures_0404.pdf; U.S. Government Accountability Office, *Critical Infrastructure Protection: Update to National Infrastructure Protection Plan Includes Increased Emphasis on Risk Management and Resilience* (Report to Members and Committees of Congress, GAO-10-296, March 2010).
34. Australian Government, Department of the Prime Minister and Cabinet, *Strong and Secure: A Strategy for Australia's National Security. Australian National Security Strategy,* January 23, 2013, http://www.nationalsecurity.gov.au
35. Dane Egli, team lead, *Facing the Storms: Operationalizing Preparedness & Critical Infrastructure Resilience* (Laurel, MD: Johns Hopkins University Applied Physics Laboratory, 2013), http://www.jhuapl.edu/ourwork/nsa/papers/FacingtheStormsFinal-1-Oct-2013.pdf

36. Thomas D. O'Rourke, "Critical Infrastructure, Interdependencies, and Resilience," *The Bridge. Linking Engineering and Society* 37, no. 1 (2007), https://www.nae.edu/File.aspx?id=7405
37. Presidential Policy Directive 8 (PPD-8), National Preparedness, March 30, 2011, http://www.dhs.gov/presidential-policy-directive-8-national-preparedness
38. Presidential Policy Directive 21 (PPD-21), Critical Infrastructure Security and Resilience, February 12, 2013, http://www.whitehouse.gov/the-press-office/2013/02/12/presidential-policy-directive-critical-infrastructure-security-and-resil
39. National Academies, Committee on Science, Engineering, and Public Policy, Committee on Increasing National Resilience to Hazards and Disasters, *Disaster Resilience: A National Imperative* (Washington, D.C.: The National Academies Press, July 2012).
40. Kerry, Lautenberg, and Gillibrand, "Strengthening the Resiliency of Our Nation."

Part II

Aspects from Disciplines

5

The Role of Intelligence in Homeland Security

WILLIAM J. RYAN

Intelligence Types and Collection Platforms

Securing the homeland is a challenging task in today's environment, and government agencies at all levels (federal, state, local, tribal, and territorial), private sector entities, and the general public play a vital role. The key mission areas of homeland security are prevention, protection, mitigation, response, and recovery. An integral ingredient in the effort to accomplish the overall homeland security mission is intelligence. Actually, the influence of accurate and timely intelligence can be seen across all of the aforementioned mission areas. Intelligence in its basic form is information that is not publicly available, is relevant, and is timely. When used in the homeland security context, intelligence can be broken down into multiple categories. For the purpose of this chapter, we will focus on the following: foreign intelligence, military intelligence, domestic intelligence, criminal intelligence, and business intelligence. Understanding the difference among these types of intelligence is integral to understanding how they can be useful:

- *Foreign intelligence:* Information relating to the capabilities, intentions, and/or activities of foreign governments or elements thereof, foreign organizations, or foreign persons, or international terrorist activities.[1] This information can be used in forming the collecting government's foreign policy, counterterrorism strategy, economic policy, etc.
- *Military intelligence:* Information collected by militaries and governments on the capabilities, plans, intentions, actions, etc., of another government's military. This information is used to form a country's military strategy.

- *Domestic intelligence:* Information collected by a government entity concerning the potentially damaging plans, intentions, actions, etc., of a group or individual within its own country. This type of information is normally used for counterterrorism purposes.
- *Criminal intelligence:* Information collected by local, state, and federal government entities in order to stop crime and affect arrests. This information is used by law enforcement agencies for all levels of criminal activity.
- *Business intelligence:* Information collected by private businesses to gain a competitive advantage over another business or to advance their standing in a particular industry.

These types of intelligence can easily cross over. For example, the collection of foreign intelligence can wind up uncovering criminal intelligence; the discovery of a planned terrorist attack is obviously the discovery of a planned criminal event; and military intelligence is also a form of foreign intelligence. However, in the United States, for example, there are laws that govern the collection of the different types of intelligence. One such law is the Foreign Intelligence Surveillance Act (FISA). The FISA requires the U.S. Government, before it commences certain kinds of intelligence gathering operations within the United States, to obtain a judicial warrant similar to that required in criminal investigations.[2] There is an exception to the FISA Act, which reads:

> Notwithstanding any other law, the President, through the Attorney General, may authorize electronic surveillance without a court order under this subchapter to acquire foreign intelligence information for periods of up to one year if the Attorney General certifies in writing under oath that –
>
> (A) The electronic surveillance is solely directed at –
>   (i) the acquisition of the contents of communications transmitted by means of communications used exclusively between or among foreign powers, as defined in section 1801(a)(1), (2), or (3) of this title; or
>   (ii) the acquisition of technical intelligence, other than the spoken communications of individuals, from property or premises under the open and exclusive control of a foreign power, as defined in section 1801(a)(1), (2), or (3) of this title;
>
> (B) There is no substantial likelihood that the surveillance will acquire the contents of any communication to which a United States person is a party; and

(C) The proposed minimization procedures with respect to such surveillance meet the definition of minimization procedures under section 1801(h) of this title.[3]

Therefore, while crossover does happen, it is usually unknown prior to the time of collection.

There are multiple collection platforms used in the acquisition of the intelligence and multiple sources from where the intelligence comes. The following is a list, along with definitions, of intelligence collection disciplines and associated collection platforms:[4]

- *Human intelligence* (HUMINT) is the collection of information from human sources. The collection may be done openly or via clandestine or covert means (espionage). Within the United States, HUMINT collection is the FBI's responsibility. Beyond U.S. borders, HUMINT is generally collected by the CIA, but also by other U.S. components abroad.
- *Signals intelligence* (SIGINT) refers to electronic transmissions that can be collected by ships, planes, ground sites, or satellites. Communications intelligence (COMINT) is a type of SIGINT and refers to the interception of communications between two parties. U.S. SIGINT satellites are designed and built by the National Reconnaissance Office, although conducting U.S. signals intelligence activities is primarily the responsibility of the National Security Agency (NSA). Authorized agencies also collect SIGINT through authorized wiretaps and other electronic intercepts of information.
- *Imagery intelligence* (IMINT) is sometimes also referred to as photo intelligence (PHOTINT). IMINT was practiced to a greater extent in World Wars I and II when both sides took photographs from airplanes. Today, the National Reconnaissance Office designs, builds, and operates imagery satellites, while the National Geospatial-Intelligence Agency is largely responsible for processing and using the imagery.
- *Measurement and signatures intelligence* (MASINT) is a relatively little-known collection discipline that concerns weapons capabilities and industrial activities. MASINT includes the advanced processing and use of data gathered from overhead and airborne IMINT and SIGINT collection systems. Telemetry intelligence (TELINT) is sometimes used to indicate data relayed by weapons during tests, while electronic intelligence (ELINT) can indicate electronic emissions picked up from modern weapons and

tracking systems. Both TELINT and ELINT can be types of SIGINT and contribute to MASINT.
- *Open-source intelligence* (OSINT) refers to a broad array of information and sources that are generally available, including information obtained from the media (newspapers, radio, television, etc.), professional and academic records (papers, conferences, professional associations, etc.), and public data (government reports, demographics, hearings, speeches, etc.).

The Intelligence Cycle

Intelligence collection is a process and is most commonly broken down into the following steps: Planning and Direction; Collection; Processing; Analysis and Production; and Dissemination.[5]

- *Planning and Direction:* In this step, intelligence professionals develop requirements that outline what information is needed to keep a country safe, formulate a country's policy, etc.
- *Collection:* This is the act of collecting the information that fulfills the requirements that were produced. As previously mentioned, there are multiple techniques used to acquire this information.
- *Processing:* The information is put into a raw intelligence report and sent to those who need to see the information pre-analysis.
- *Analysis and Production:* Analysts put the information in context, often taking bits and pieces from multiple raw intelligence reports and put the corroborated information into a finished analytical piece for dissemination.
- *Dissemination:* The analyzed information is sent to those who need the information to make decisions. Those recipients use the information, and provide feedback outlining additional information needs from which additional requirements are formulated, starting the cycle over.

This cycle is practiced in some similar form by all of the members of the "intelligence community." The members of the intelligence community all have their respective roles in contributing to the protection of the homeland. The members of the federal intelligence community in the United States are:

- Air Force Intelligence;
- Army Intelligence;
- Central Intelligence Agency;

- Coast Guard Intelligence;
- Defense Intelligence Agency;
- Department of Energy;
- Department of Homeland Security;
- Department of State;
- Department of the Treasury;
- Drug Enforcement Administration;
- Federal Bureau of Investigation;
- Marine Corps Intelligence;
- National Geospatial-Intelligence Agency;
- National Reconnaissance Office;
- National Security Agency; and
- Navy Intelligence.[6]

While the aforementioned agencies make up the U.S. intelligence community, other countries around the world have their own intelligence agencies. Because of the proliferation of global intelligence collection, many countries practice counterintelligence. The term "counterintelligence," as noted in 50 USC 401a, means

> information gathered, and activities conducted, to protect against espionage, other intelligence activities, sabotage, or assassinations conducted by or on behalf of foreign governments or elements thereof, foreign organizations, or foreign persons, or international terrorist activities.[7]

So, while governments go to great lengths to collect the aforementioned types of intelligence, they also recognize the importance of preventing, to the extent possible, collection efforts against their homelands.

Intelligence Classification, Controls, and Dissemination

The collected and disseminated intelligence is given a classification. A classification is what dictates how widely or narrowly the information is shared. Often times, intelligence is classified to whatever level is needed to protect the source of the information—regardless of whether the source is human or electronic. In the United States, there are three levels of classified material:

- *TOP SECRET* information is information which, if disclosed without authorization, could reasonably be expected to cause exceptionally grave damage to the national security.

- *SECRET* information is information which, if disclosed without authorization, could reasonably be expected to cause serious damage to the national security.
- *CONFIDENTIAL* information is information which, if disclosed without authorization, could reasonably be expected to cause damage to the national security.[8]

There are also additional controls that can be added to the classification of information in order to further restrict its dissemination. While the above classifications are used by governmental bodies in the United States, all countries have a system to restrict the dissemination of sensitive information. Mirroring the governments' efforts to restrict sensitive information sharing, the private sector, comprised of companies that play a vital role in protecting the homeland, also have controls on sensitive internal information. When this information is shared with government entities, it is usually with the understanding that the government will protect the information the same way the company would.

Consumers of Intelligence and the Criticality of Actionable Intelligence

Most local, state, and federal government agencies are consumers of intelligence. They receive the types of intelligence that assist them in completing their missions. As stated previously, governments also formulate policy based on intelligence. Most federal governmental agencies have counterparts in other countries around the world. Oftentimes, countries will cooperate and share intelligence when it is to their mutual benefit—this can be seen often with counterterrorism efforts. Additionally, the private sector is a consumer of intelligence, which enables a private entity or entities to better protect assets from a known threat, or to assist the government in its collection efforts.

When a piece of intelligence needs to be widely disseminated and the intended audience does not have a security clearance, intelligence can then be re-written to allow for the removal of the classification and controls, therefore allowing it to be disseminated further, sometimes to the general public. This has happened many times in the past, specifically with threat information. There are times when the general public needs to be informed about a potential threat not only to take potential action to protect themselves and their loved ones, but also to be another set of eyes and ears for those governmental agencies attempting to prevent the threat from coming to fruition. Prevention would come in the form of either stopping the potential perpetrator (via arrest or other means) or making it so the intended targets are no longer vulnerable to the planned action.

As previously mentioned, many actions taken to secure the homeland are based on intelligence or can be enhanced with accurate and timely intelligence. When you look at all of the efforts undertaken to protect the homeland, they can range from: protecting the border from being infiltrated by those wishing to do harm; protecting a nation's critical infrastructure from potential damage (physical and cyber); tracking the shipment of cargo from the country of origin to final destination to ensure nothing untoward was introduced into the supply chain; monitoring restricted technology or products from a country so it does not fall into the wrong hands, whether accidentally or with intent; ensuring the financial health of a country's economy; securing a country's assets that reside outside of its geographic footprint; etc. Intelligence allows those responsible for these efforts to better focus their resources to best carry out a specific mission.

Let us look at a hypothetical scenario where actionable intelligence is collected, analyzed, and disseminated:

> *A human source located in country X provides information to a collector from country Z that a group of individuals from country X are travelling via country Y to arrive in country Z to conduct an attack against a piece of critical infrastructure that will cause great loss of life and economic damage. Materials to be used in the attack will be purchased in countries Y and Z, based on how the group views the security, or lack thereof, at points of purchase. The human source's byline is one that gives confidence to the collector and recipients of the intelligence that the information is relatively accurate.*

This information is written up by the collector (or collecting agency), checked for source protection, classified, and disseminated in raw form to those that need to see it immediately. The raw intelligence will include some version of a source byline, which will give those who eventually receive the intelligence an idea of its source's reliability, therefore speaking to the authenticity and reliability of the intelligence. Those receiving the raw (not yet analyzed and put into a finished piece, as described earlier) intelligence will make decisions, in coordination with their partners who share responsibility for protecting the homeland, on actions to be taken based on the intelligence. In this instance, some or all of the following actions may be taken:

- *Country Z will share the intelligence with country Y, via its liaison relationship, so country Y can be on alert for these individuals entering its country.*

This intelligence would likely be shared in a way as to protect its original source—the human source in country X.

- *Assuming the names of the travelers are known, their names and any known aliases would be placed on a watch list at border crossings (land, sea, and air) in order to stop them from entering into either country Y or Z.*
- *General alerts could be disseminated by the government to national and local (if specific geographic areas were mentioned in the intelligence) retailers of raw materials that could be used to construct explosives or other harmful weapons, asking them to call their respective government emergency numbers if they see anything out of the ordinary.*
- *If anything is known about the potential targets, the government could either disseminate a general alert if only the critical infrastructure sector(s) is known, and/or brief specific facilities if the intelligence was more specific about the target(s).*
- *A review of the security of either the critical infrastructure sector in general or for specific targets could take place between the government and the owners and operators of the potential critical infrastructure targets.*
- *The private sector owners and operators would likely increase the security at the potentially targeted facilities.*
- *Government law enforcement and security personnel could be deployed in the potential attack area(s) as a show of force and to increase security posture.*
- *During this entire process, government law enforcement, intelligence, and security personnel are trying to locate the potential perpetrators of the planned attack.*
- *Eventually, the raw intelligence is analyzed, combined with other information that has been collected, and put into a classified finished piece of intelligence, disseminated within the intelligence community, and to other authorized recipients to give a more in-depth and sometimes broader view of the situation; however, as was seen above, the raw intelligence was actionable and needed to be addressed immediately.*

As can be seen from this example, the intelligence collected could mobilize an array of homeland security stakeholders to carry out their specific missions with one goal in mind: to stop the attack from happening or, in the worst case, mitigate the effects of the attack if it were somehow perpetrated.

Partnerships

As has been mentioned, there are many different intelligence collectors on the local, tribal, territorial, state, and federal levels—as well as a lot of useful information residing in the private sector. The collection, dissemination, and use of intelligence, vital in protecting the homeland, can only be accomplished fully via partnerships. Realizing that the private sector plays a significant role in this partnership, there are agencies within the federal government in the United States that grant security clearances to key homeland security stakeholders with certain companies, organizations, and industries in order to assist the government in obtaining intelligence to protect the homeland, but more often to allow the government to more easily disseminate classified intelligence to these stakeholders. The dissemination of this intelligence allows the private sector to act on the information to protect their vital assets, and assist the government in determining what actions should be taken to prevent and/or mitigate the potential effects of an individual, organization, or nation-state trying to do harm to their specific company or industry in general. Additionally, based on intelligence collected globally, many counties educate their private sector partners through formal and online courses to increase private sector security posture and awareness. This education goes both ways as private sector partners provide valuable insight as to the inner workings of their specific sectors and/or industries and what is needed to secure their assets. The cooperation also extends to tabletop, functional, and full-scale exercises to practice and more fully understand the scope of what is being verbally communicated.

Any successful homeland security program has to include public awareness campaigns. This would include not only educating the general public as to the threats that could affect their security, safety, and/or way of life, but also how members of the public could report suspicious activity or behavior to assist in preventing a man-made incident. Consider the *homegrown violent extremists* (HVEs), for example. According to the U.S. Department of Homeland Security, using a White House definition, violent extremists are defined as "individuals who support or commit ideologically-motivated violence to further political goals."[9] Violent extremist threats within the United States can come from a range of violent extremist groups and individuals, including domestic terrorists and homegrown violent extremists. The U.S. Department of Homeland Security defines *domestic terrorism* as:

> Any act of violence that is dangerous to human life or potentially destructive of critical infrastructure or key resources committed by a group or individual based and operating entirely within the United States or its territories without direction or inspiration from a foreign terrorist group.[10]

Contributing to the challenges associated with detecting the potential actions of HVEs is that some do not have criminal records nor have they been on the radar of law enforcement agencies. How then can we discover the plans and/or intentions of these HVEs? One very effective way is via tips from the general public. No matter in which country the HVE resides, there are generally signs that are present in and around the familial, workplace, educational, and/or social circle that would cause someone concern. These instances need to be reported to an official representing public safety and security in some capacity in order to stop the HVE before loss of life and/or damage to public property occurs. Without this type of public cooperation, the collection of this domestic intelligence becomes exponentially more difficult.

Another effort that has grown out of the need for partnerships is the creation of intelligence fusion centers. In the United States, there is a fusion center in every state, along with several regional centers (often in larger, urban areas) that disseminate information regionally and also feed the state centers. A fusion center is a location (usually physical but can be virtual as well) where officials from local, state, and federal governments, often along with members from the private sector, work alongside each other to ensure any pertinent intelligence is being shared across the spectrum of those with responsibilities for securing the homeland. Each organization represented in the fusion center has its own needs in terms of workspace, database setup, and protection of the information its agency/organization produces—but everyone with access to the intelligence that is shared within the fusion center has a security clearance to the appropriate level. After sharing the intelligence with each other within the fusion center, the intelligence is analyzed, put into a product, and disseminated to those in the public and private sector homeland security world that need to receive it. This process follows the intelligence cycle described earlier in this chapter. To better understand this dynamic, consider the following example:

> *A U.S. Government entity intercepts the electronic communications of a suspected terrorist, located within the U.S., which specifies both physical attacks against government buildings and cyber attacks against private industry set to take place in multiple named cities around the United States and in Europe. The attacks will attempt to take advantage of physical and cyber vulnerabilities within government and private sector agencies. The suspected terrorist also mentions timeframes.*

On the U.S. side, the collecting agency disseminates the intelligence to those with a need to know and appropriate clearances at the federal level. The intelligence is then briefed at the fusion centers in the aforementioned cities to the state and local (and tribal and territorial if applicable) officials, again with the need to know and appropriate clearances. Local and state law enforcement and security agencies in the potential target cities comb through their databases trying to find unusual activity that has taken place over a specified period of time (including traffic stops, local arrests, suspicious activity reporting, etc.) and share that information with the federal entities at the fusion center. Private sector fusion center partners in industries heavily reliant on cyber are asked to check and double-check their systems for recent unusual activity. Federal, state, and local entities will review security/vulnerability assessments that have been conducted on high-profile facilities in their respective areas.

The fusion centers in the named cities share the information they have collected with each other to get a better understanding of what may be planned. Local, state, federal, and private sector entities can then come together with all of the available information to map out a plausible plan to prevent the attacks from taking place—or at least mitigate the potential effects. The concept of fusion centers is to make all of this sharing easier and more efficient.

While the fusion center activity is taking place, U.S. Government officials with foreign government liaison duties are briefing their counterparts in the named European cities so the cities' specific capabilities can be brought to bear against the threat.

As is shown by the simplified example above, partnerships are an absolute necessity. Through these partnerships, fusion centers have supported efforts as varied as disrupting plots against the homeland, combating transnational drug networks, and apprehending fugitives; and while the aforementioned example specifies potential terrorist activity, negative impacts resulting from natural events are also a great concern. Because of this, many fusion centers have made the effort to become all-hazards centers. An all-hazards center focuses not only on man-made threats to the homeland, but also on natural threats, such as hurricanes, earthquakes, or tornadoes. Natural disasters have proven to be just as damaging, if not more so, than man-made incidents. An all-hazards fusion center would include members of a locality or region's fire service, department of emergency management, etc. For example, fusion centers played a role during Superstorm Sandy (2012) by providing geospatial support, damage reporting, and criminal/terroristic threat reporting associated with those looking to take advantage of the situation, etc.

Challenges

As has been alluded to in this chapter, there is a large amount of information collected by local and state agencies/departments that makes its way into intelligence reports produced at the federal level. There is also information that is collected at state and local levels that is more criminal in nature and does not meet the requirements to be included in a federal intelligence report. Understanding the difference between the two types of information can be a challenge to collectors, especially for those in law enforcement that have spent a career collecting information for the purpose of affecting an arrest. Law enforcement officers are very good at doing what they have been trained to do, but world events have forced new roles and responsibilities on them. Understanding that an arrest they make, or information that is gathered at a crime scene, can be vital to uncovering a larger terrorist or nation-state counterintelligence plan or plot, has definitely taken hold with many law enforcement agencies. However, analyzing that information and knowing how to see it in a broader context can only come through training and experience.

On the opposite end of the spectrum, intelligence collectors and analysts that have been trained to collect and analyze information in a broader context have an easier time seeing the larger picture. This becomes instinctive as one's experience deepens. But when criminal elements appear in the intelligence collected, a law enforcement entity will then weigh in and a decision is made on how to handle the situation. The bottom line is that you can't take a collector or analyst that sees the information/intelligence through one lens and expect him or her to be able to adjust to doing business another way without the proper training.

Another challenge with which homeland security stakeholders are faced is the proliferation of undercover operations at all levels of government. Local, state, and federal law enforcement agencies, as well as federal intelligence agencies, run undercover operations for various reasons to accomplish their respective missions. The practice of running undercover operations on various levels is not isolated to the United States, but happens on an international scale as well. This practice further demonstrates the need to cooperate and coordinate both domestically and internationally. One method of coordination on the counterterrorism front that occurs in many jurisdictions in the United States is via Joint Terrorism Task Forces (JTTF). JTTFs are coordinated under the auspices of the Federal Bureau of Investigation. JTTFs are made up of members from local, state, and federal law enforcement, security, or intelligence agencies. The members of the JTTF work together for counterterrorism purposes.

As previously mentioned, on the international front intelligence is often shared among nations when it is to the benefit of both countries, but governments are at times hesitant to share too much as they do not want to reveal their methods for obtaining the intelligence or the source of the intelligence in question. Intelligence agencies for respective countries will often employ liaison officers to interact with a foreign government for the purpose of cooperation and intelligence sharing.

The same practice occurs within a country. For instance, within the United States, agencies and/or departments share information via fusion centers or other means, but are often careful not to divulge a plethora of operational details, so as not to put their specific operation or investigation at unnecessary risk. This is when many follow the "need to know" policy—even though someone may have a clearance to know a certain piece of information, if they do not have a specific need to know, then it need not be shared.

One can plainly see the challenges that come along with making decisions on how much intelligence and/or operational information to share, as something left out of an exchange could potentially affect the efforts of another organization or department within a country, or efforts of an international partner.

Securing the Homeland

The 2014 Quadrennial Homeland Security Review (QHSR) stated that the United States should continue to adhere to the five basic homeland security missions as already set forth in the first Quadrennial Homeland Security Review report in 2010. These mission areas are:

- "Prevent Terrorism and Enhance Security;"
- "Secure and Manage Our Borders;"
- "Enforce and Administer Our Immigration Laws;"
- "Safeguard and Secure Cyberspace;" and
- "Strengthen National Preparedness and Resilience."[11]

QHSR 2014 goes on to state that to set homeland security priorities, the U.S. Department of Homeland Security leads national efforts to assess, analyze, and compare risk, which is a function of the likelihood and potential impacts of different homeland security threats and hazards. QHSR 2014 recognizes that the likelihood and consequence of specific threats and hazards may be influenced over time by interdependent economic, political, social, environmental, and technological factors, as well as trends and future uncertainties.

James Steiner, in his book entitled *Homeland Security Intelligence,*[12] relates in detail how the intelligence apparatus in the United States works to accomplish the missions outlined in QHSR 2014. Homeland Security is an all-encompassing effort in which everyone plays a part: local, state, tribal, territorial, and federal governments; private sector entities; and the general public. By collecting, disseminating (appropriately), and protecting intelligence, countries around the world can be better prepared for the aforementioned uncertainties that could, and undoubtedly will, affect their homeland security.

Notes

1. *U.S. Code (2006), Supplement 5, Title 50—WAR AND NATIONAL DEFENSE,* Section 401a—Definitions, http://www.gpo.gov/fdsys/granule/USCODE-2011-title50/USCODE-2011-title50-chap15-sec401a
2. "Foreign Intelligence Surveillance Court," Federal Judicial Center, last accessed February 23, 2014, http://www.fjc.gov/history/home.nsf/page/courts_special_fisc.html
3. *U.S. Code (2006), Supplement 5, Title 50,* Section 401a.
4. "Intelligence Collection Disciplines (INTs)," Federal Bureau of Investigation, last accessed February 23, 2015, http://www.fbi.gov/about-us/intelligence/disciplines
5. "The Intelligence Cycle," Central Intelligence Agency, last updated March 23, 2013, https://www.cia.gov/kids-page/6-12th-grade/who-we-are-what-we-do/the-intelligence-cycle.html
6. "Members of the IC," Office of the Director of National Intelligence, last accessed February 23, 2013, http://www.dni.gov/index.php/intelligence-community/members-of-the-ic
7. *U.S. Code (2006), Supplement 5, Title 50,* Section 401a.
8. Executive Order 13526 of December 29, 2009, Classified National Security Information. *Code of Federal Regulations*, Title 3 (2010): 298–327, http://www.gpo.gov/fdsys/pkg/CFR-2010-title3-vol1/pdf/CFR-2010-title3-vol1-eo13526.pdf
9. "Countering Violent Extremism," U.S. Department of Homeland Security, last published October 15, 2014, http://www.dhs.gov/topic/countering-violent-extremism
10. Ibid.
11. U.S. Department of Homeland Security, *The 2014 Quadrennial Homeland Security Review* (Washington, D.C., June 18, 2015), http://www.dhs.gov/sites/default/files/publications/2014-qhsr-final-508.pdf
12. James E. Steiner, *Homeland Security Intelligence* (Thousand Oaks, CA: CQ Press, 2015).

6

Geospatial Intelligence and the Geospatial Revolution

TODD S. BACASTOW & GREGORY A. THOMAS

Introduction

This chapter examines how the revolution in geospatial technology combined with the tradecraft of geospatial intelligence (GEOINT) has changed how analysts develop insights about how humans use geography. Advances in satellites, global positioning systems (GPS), unmanned aerial systems, wireless communications, handheld computing, and the ability to automate laborious map analysis processes have transformed what used to be called geographic intelligence, and the nature of the insights provided to managers and leaders. GEOINT helps daily with real-time applications to guide decision-making.

Five principles that serve as a foundation for geospatial intelligence will be addressed throughout this chapter. These principles are:

- *Principle #1:* Geospatial intelligence concerns insights about human activity on the Earth.
- *Principle #2:* The analyst's craft completes the picture provided by geospatial intelligence data.
- *Principle #3:* Geospatial intelligence collection and source strategies are focused on answering questions about place and time.
- *Principle #4:* The geospatial intelligence analysis tradecraft is the art of geospatial reasoning informed by geographic information science aided by geographic information technology.
- *Principle #5:* The geospatial intelligence analyst makes judgments consistent with the standards of the intelligence profession.

GEOINT is defined by the intersection of the intelligence tradecraft, geographic science, and geographic information technology. As a discipline,

GEOINT provides decision makers with insights about the ways in which people organize and arrange their activities on Earth's surface. These insights can prevent surprises, leverage emerging opportunities, counteract threats, or provide time to adapt to changing situations. GEOINT provides decision makers with actionable knowledge not only in the national security sector, but also in business, humanitarian efforts, law enforcement, and homeland security.

Because homeland security covers such a broad array of issues from transportation, border control, emergency management, and response, health monitoring, infrastructure protection, intelligence gathering, and law enforcement, geospatial intelligence has many applications in this field. Homeland security encompasses an all-hazards approach to threats posed not only by terrorist attacks, but also by natural disasters. Homeland security geospatial intelligence involves the analysis of human events on Earth over place and time to prevent, protect against, mitigate, respond to, and recover from terrorist incidents and other man-made or natural disasters. For this reason, geospatial intelligence is an important component to homeland security studies. It is a valuable skill set for the analysis of information. The following provides an overview of the principles of geospatial intelligence and demonstrates its broad application.

What Is Geospatial Intelligence?

To explain what is unique about this discipline that distinguishes it from other branches of geography and geographic analysis, it is important to first define the individual terms *geospatial* and *intelligence*.

- *Geospatial* pertains to or relates to the relative position of things on the Earth's surface.
- *Intelligence* is information gathered to guide decisions. It includes information that may be both open and closed, and obtained from many different sources.

Expanding upon the word *geospatial*, while the words geospatial, geographic, and spatial are often used interchangeably to mean similar things, the reasoning behind the linguistic blend forming geospatial is that spatial alone is too generic and geographic is too related to the particular discipline of geographic intelligence,[1] one of the oldest forms of military intelligence.

Expanding upon the word *intelligence*, intelligence provides a decision advantage intended to prevent surprise, capitalize on emerging opportunities, neutralize threats, or provide time to adapt to a changing situation. Intelligence is often associated with secrets and spying. However, according to experienced national security expert Mark Lowenthal, viewing intelligence as primarily secrets misses the important point that intelligence is ultimately information that meets the needs of a decision maker. Intelligence is information that has been collected, processed, narrowed, and offered to meet those needs, which could apply to any domain. Lowenthal also asserts a central notion that intelligence is not about truth, but is more accurately thought of as a "proximate reality."[2] So, semantically, GEOINT provides insights to a decision maker about how humans relate to the Earth.

The Official Definition of GEOINT

The official United States Government definition of *geospatial intelligence* is stated in the U.S. Code (10 USC 467) as follows:

> The term "geospatial intelligence" means the exploitation and analysis of imagery and geospatial information to describe, assess, and visually depict physical features and geographically referenced activities on the earth. Geospatial intelligence consists of imagery, imagery intelligence, and geospatial information.[3]

It not unreasonable to say that this definition was crafted to inform how the National Geospatial-Intelligence Agency (NGA) operated with the other U.S. intelligence agencies, but was not intended as the functional definition of the discipline. In the U.S., the term GEOINT is used to connote a source of intelligence like human intelligence (HUMINT)[4] or signals intelligence (SIGINT)[5]. The national differences notwithstanding, GEOINT is different from the other sources (e.g., SIGINT) and is inherently multi-source. It integrates and enriches information collected by the other intelligence collection disciplines (INTs) into a space-time context.

Emerging Definition of GEOINT

The narrow definition of GEOINT in the U.S. Code does a disservice to the discipline when considering geospatial intelligence is broader than as narrowly defined in U.S. law. While geospatial intelligence is essential in the traditional intelligence areas of deterring war, resolving conflict, and promoting peace, GEOINT is also important to other areas such as supporting the commercial sector and civil authorities. There is an emerging definition of geospatial intelligence that is vastly different from the U.S.'s statutory definition. Geospatial intelligence has evolved to a more encompassing definition of:

> Geospatial intelligence is actionable knowledge, a process, and a profession. It is the ability to describe, understand, and interpret so as to anticipate the human impact of an event or action within a place-time environment. It is also the ability to identify, collect, store, and manipulate data to create geospatial knowledge through critical thinking, geospatial reasoning, and analytical techniques. Finally, it is the ability to present knowledge in a way that is appropriate to the decision-making environment.[6]

Central to this definition is the notion that the focus of GEOINT is understanding how people organize and arrange their activities on Earth's surface. For all the emphasis in GEOINT on data and technologies, people remain GEOINT's most precious analytic resource. The analyst describes, understands, and interprets so as to anticipate the human impact of an event or action. The analyst's work involves the analysis of *place* and *time* using geographic information sciences, geographic technologies (e.g., satellites, unmanned aircraft systems, GPS, geographic information systems, and cyber) and a unique tradecraft.

How Is GEOINT Different?

GEOINT is a sub-discipline of geography and is unique from other forms of geospatial analysis because of its tradecraft. Specifically, GEOINT delivers insights gained from *place* and *time for* a decision advantage *by* integrating *geographic information science*, *geographic information technology*,[7] and its unique *tradecraft*.

Place

Place is fundamental to geography and perhaps the most important GEOINT concept. At first glance, location and place might seem to be similar terms. However, *places* have physical and human attributes that make them what they are. Physical attributes may include a description of such things as the mountains, rivers, beaches, and topography of a place. Human characteristics may include the human-designed cultural features of a place, from land use and architecture, to forms of livelihood and religion, to food and folkways, to transportation and communication networks. *Place* emphasizes the understanding of both of these factors and their integration.

Ultimately, GEOINT is concerned with understanding why places, and people in those places, are located where they are and how they interact. To answer this question, the analyst must be comfortable with the underlying concepts and theories of the spatial distribution of a particular phenomenon.

Time

Geography and, hence, GEOINT, approaches the explanation of spatial phenomena through the change of geographical features. Over 40 years ago, a Swedish geographer, Torsten Hägerstrand, illustrated how human spatial activity is often governed by time limitations. He identified three categories of limitations: capability constraints, coupling constraints, and authority constraints.[8]

- Capability constraints refer to the limitations on human movement due to physical or biological factors. For example, a person cannot be in two places at one time.
- A coupling constraint refers to the need to be in one particular place for a given length of time, often in interaction with other people. For example, your space-time path must temporarily link up with certain people to accomplish a particular task.
- The authority constraint is a controlled area that sets limit on its access to particular individuals. For example, a person's space-time path is normally not permitted to enter a sensitive military base.

Hägerstrand's concept of space-time is powerful. This model provides a framework for understanding human activity,[9] and it provides a theoretical foundation for intelligence concepts such as activity-based intelligence.[10]

Geographic Information Science (GIScience)

Geographic information science (GIScience or GISci) includes scientific research areas of geographic information systems (GIS), cartography, remote sensing, photogrammetry, and surveying. GIScience addresses fundamental issues in the use of digital technology to handle geographic information about places, activities, and phenomena on and near the surface of the Earth that are stored in maps or images. GIScience includes fundamental questions of data structures, analysis, accuracy, meaning, cognition, and visualization to name a few. As such, GIScience overlaps with many traditional disciplines such as earth science, mathematics, computer science, physics, cognitive science, and ethics. While GIScience is not central to any of these disciplines, researchers from many different backgrounds work together on particular sets of interrelated problems.

Geographic Information Technologies (GIT)

The term *geographic information systems* (GIS) has been used to describe the hardware, software, geographic data, and people that enable geospatial information processes for making decisions. More recently, GIS, together with global positioning systems, remote sensing techniques, and other spatially related tools for decision-making, comprise a larger array of tools that can be grouped together under the more complete term of *geographic information technologies* (GIT).

Tradecraft

Tradecraft is what makes GEOINT unique. It encompasses the unique organizational sources and methods for obtaining GEOINT data and making sense of it to support the decision maker.[11] GEOINT's methods comprise the tools to organize geospatial data and the reasoning techniques for rendering judgments, insights, and forecasts about human activities and intentions. The skills of detecting geospatial deception and, where required, maintaining secrecy of geospatial sources and methods makes GEOINT unique. Secrecy, which is often contrasted with transparency as an ideal, has negative connotations and is often associated with spying and espionage. However, in a real way, total transparency is unnatural and seldom occurs. Humans conceal aspects of their lives from others due to fear of inappropriate use of the information, embarrassment, retribution, denunciation, harassment, or loss of employment.

Organizing Data for Prediction

Geospatial Intelligence Data

Geospatial information, imagery, and imagery intelligence are the primary sources from which GEOINT data are extracted. However, there is a more foundational perspective. These data are records of human and physical qualities at a location that uniquely created a place. Signs of human activity and measures of the land become GEOINT data. Traditionally data are thought of as fixed statements of "what is"; however, GEOINT data are part of a conversation driven by the need to explain what is recorded in the data in the context of a place. GEOINT data are extracted chunks of information from other intelligence and geographic products. Inherent in using this data, GEOINT data require knowledge of the data and how the spatial features might be organized on the landscape. Using the data and understanding spatial organization allows analysts to frame patterns of human activities to make predictions. GEOINT data fitted into a frame also guide the search for additional information and ultimately provide a plausible story or map to account for what the data show.

Past National Geospatial-Intelligence Agency doctrine said that GEOINT data "is any data used to create GEOINT."[12] GEOINT data are data and information collected to understand the human and physical qualities of a location. GEOINT data fundamentally consist of structured and unstructured data, and information about the physical and human qualities of a location on the Earth. Table 6.1 illustrates these general data types and the relation between them:

Table 6.1. Data Content vs. Data Organization.

	Data Organization: Structured	Data Organization: Unstructured
Data Content: Physical Geography	Example: Digital imagery of a land feature	Example: Report describing a land feature
Data Content: Human Geography	Example: Geospatial data of incidences for bacterial infections	Example: Scholarly article describing conditions associated with bacterial infections

Structured GEOINT data are geospatial data organized to be immediately usable by technologies, such as geographic information systems (GIS). A formal definition is:

> Structured geospatial data is information about locations and shapes of geographic features and the relationships between them. It is usually stored as coordinates and topology with a high degree of organization to be readily searchable.

What often goes unstated and unappreciated is a broad class of data known as "unstructured geospatial data," a catch-all term because much of the data included under that term actually have elements of structure. Email, for instance, may contain a street address, senders, times, and the like. A definition for unstructured geospatial data is:

> Unstructured geospatial data refers to geographic information that either does not have a pre-defined data model or is not organized in a pre-defined manner. Unstructured geospatial data may be text containing geographic information such as street addresses and site descriptions. Unstructured data is not readily searchable.

GEOINT data can be divided into two other major data content categories of physical and human geospatial data. Physical geospatial data is a record of the spatial characteristics of the various natural phenomena associated with the Earth's hydrosphere, biosphere, atmosphere, and lithosphere. Examples are:

- landforms that are a result of the natural processes;
- vegetation; and
- climate and weather.

Human geography data record the imprint of human activity on Earth. Examples include:

- cities, buildings, roads, bridges, power/utility lines, and airfields;
- farms, orchards, irrigation channels, deforestation, dams, and mining; and
- birth and death rates, population clusters, media accessibility, political culture, medical facilities, education, tribal boundaries, and other cultural details.

GEOINT data is not really data in the strict definition of the term. GEOINT data are extracted chunks of information from other intelligence and geographic products such as imagery and maps.

Satellite imagery is one of the key ingredients of GEOINT. A satellite image is processed data. A satellite image is made up of raw data (thousands of pixels) that are arranged in a particular way. The raw data are processed, organized, structured, and presented so as to make them useful as a satellite image. Importantly, this satellite imagery can then be used to create other data. For example, the height of a particular building extracted from the imagery is new data.

The following Figure 6.1 illustrates this data-information-insights process.

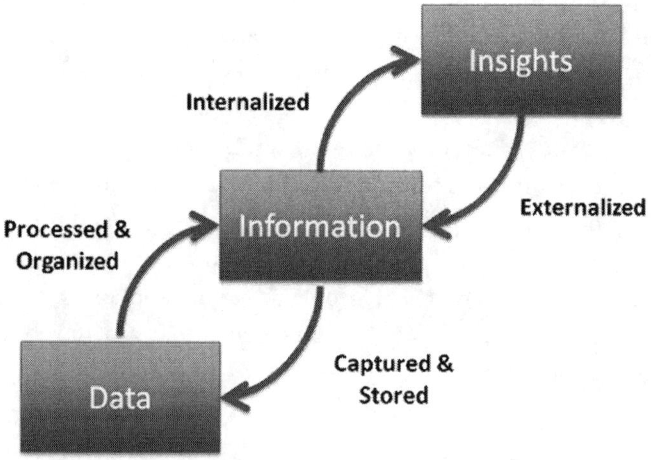

Figure 6.1. Data-Information-Insights Process.
Source: Todd S. Bacastow.

Bridging the Gap between Data, Information, and Insights

Models, both mental and otherwise, play a critical role in bridging the gap between data, information, and insights. They are an idealized representation of how to use data to solve problems. Models represent an understanding of how the world works. Analysts construct models for:

- simplification and organization;
- communication with oneself or others;
- prediction; and
- manipulation.

Models also give the ability to overcome incomplete data by mentally filling in gaps, making an intuitive leap with only the sparsest of data. This is a sophisticated form of geospatial reasoning. Expertise in geospatial reasoning increases with experience because as analysts learn or experience additional models, their minds expand to accommodate them.

An individual can "preload" their mental models with typical understandings of place; these include conceptual models of how a place is organized and works. For example, why individuals use a store at a particular location is related to the number of people in the surrounding community. Theories of spatial organization can be a shortcut in an attempt to model an unfamiliar pattern. Mental models are a normal, everyday human activity essential for geographic problem solving.

However, a word of caution is in order. While models provide a useful way of understanding the world around us, blind adherence to a model can be disastrous. When analysts close their minds to disconfirming evidence and the possibility of alternative outcomes, they fail to see the weaknesses of the model and will fail. History is replete with examples of people adhering stubbornly to an outdated paradigm (model) despite overwhelming evidence that a new way of thinking (a new model) is necessary.

Models, Frames, and Insights

Models and theories provide a basis for a structure that is called a frame. A frame helps analysts understand the world and may be influenced by a model or theory. Models and theories may also direct attention toward the information sought. In summary, individuals frame things to make sense of what they see in the real world and to fill in missing data. The frame may be influenced by an existing model or theory. A frame can take the following forms:

- A narrative explaining the chronology of events and the causal relationships between them.
- A map showing distances, directions, and connections of things such as landmarks, routes, and destinations.

Insights result from a narrative that accounts for what the data reveals within the frame; this human process of explanation is termed sense-making. There are three primary outcomes for GEOINT sense-making:

1. To identify patterns.
2. To describe patterns.
3. To predict future patterns.

GEOINT data is a key part of a discourse to make sense of indefinite and ambitious situations. Fitting the GEOINT data to the frame involves cognitive work to understand the relationships amid data and sequence of events.

GEOINT Data

This section introduces GEOINT data sources and collection strategies and focuses on the third GEOINT principle that geospatial intelligence source and collection strategies are focused on answering questions about place and time. GEOINT data are collected from a variety of sources using various collection strategies. In general, "sources" are the means or systems used to observe and record data. The term "collection strategy" is used to convey a high-level approach to GEOINT data collection. The nature of access to the data is important and determines the source.

GEOINT Data Requirements, Sources, and Collection Strategies

The purpose of GEOINT is to supply decision makers with timely geospatial insights that allow for informed, knowledgeable decision-making. In order to fulfill this purpose, analysts prioritize the intelligence requirements of the decision maker. These requirements define the mission, functions, and structure of the GEOINT; they also should drive GEOINT data collection and analysis. The Intelligence Cycle starts with requirements.[13] The requirements are sorted and prioritized, then used to drive the collection activities. Once information has been collected, it is initially evaluated, processed, and reported to consumers. Based on the requirements, collection systems are given specific tasks to execute. Requirements are often difficult to define because they require a focus on long-term needs while most decision makers do not know what information they want until they actually need it.

Collection includes acquiring information and providing that information for processing and production. Collection management is the formal process within an intelligence organization of converting intelligence requirements into collection requirements, establishing priorities, tasking, coordinating with the collection sources, monitoring results, and re-tasking as required. The collection process encompasses the management of various activities, including developing collection guidelines that ensure the best use of resources. There are four management criteria:

- Is this intelligence necessary?
- Is it feasible to collect this information?
- Is it timely?
- Is this intelligence requirement sufficiently specific?

A collection strategy seeks to determine if sources can satisfy the requirements. Three key goals of a collection strategy are to:

- meet the information need;
- collect information to detect deception; and
- provide redundancy in the event a source is unavailable.

Detecting deception requires information from a variety of sources so that one source can be verified. Multiple collection sources enable collection managers to cross-cue between different sources. Collection may require redundancy so that the loss or failure of one source can be compensated by another source.

GEOINT data are collected by computer systems, automated sensors, and humans. In general, intelligence sources are the means used to observe and record information relating to the condition, situation, or activities of a targeted location, organization, or individual. GEOINT data sources have traditionally included imagery and geospatial data. While imagery is still the dominant and most important source, GEOINT is evolving to integrate forms of intelligence and information beyond the traditional sources of geospatial information. This section recognizes this transition and structures the discussion in broad terms.

GEOINT data sources can be categorized from discontinuous to persistent versus closed to open, shown in Table 6.2 below.

Table 6.2. GEOINT Data Sources and Strategies.

	Collection Strategy: Discontinuous	Collection Strategy: Persistent
Source Access: Open	Example: The home location of the students taking a course	Example: A continuously monitored video camera along a street
Source Access: Closed	Example: The Coca-Cola Company's Coca-Cola recipe	Example: UAV equipped with a camera tracking a military target for 24 hours

A strategy is a plan to achieve goals; it provides direction and scope for an effort. In this section, the phrase "collection strategy" is a way to pursue GEOINT data collection. A collection strategy is important because the resources available to achieve goals are usually limited. For discussion purposes, this section divides a continuum of strategies into two categories of persistent collection and discontinuous collection. The collection strategy determines the frequency at which data are captured for a specific place on the Earth. This frequency is termed temporal resolution. The more frequently data are captured by a particular sensor, the better or finer is the temporal resolution of that sensor. Temporal resolution is relevant when using imagery or elevation datasets captured successively over time to detect changes to the landscape.

Discontinuous collection is not a full record of activity during a time period. It also might be termed non-persistent. Discontinuous collection is not a "permanent stare" at a target from one or more systems. Such a permanent stare might not be available, technically feasible, or an efficient use of resources. The practice of discontinuous data collection currently dominates the discipline.

Traditional remote sensing is an example of discontinuous source strategy. Here, sensors are mounted on board aircraft or spacecraft orbiting the Earth. At present, there are several remote sensing satellites providing imagery for research and operational applications. Space-borne remote sensing provides repetitive coverage of an area at relatively low cost per unit area. Discontinuous collection can also occur when a device sporadically provides data. For example, a roaming cell phone locates itself with signals from multiple antennas or through its GPS. There are numerous other examples of discontinuous GEOINT data. For instance, toll highway devices record point of entry and point of exit to compute a toll. Twitter is another example. Tweets can be geotagged to record information about the location of the device that created the tweet. Images taken with a phone or a GPS-enabled camera can also contain location data.

Persistent collection is defined as a strategy that emphasizes the ability to linger for a period of time to detect, locate, characterize, identify, track, target, and possibly provide near- or real-time data. Persistent collection facilitates the collection and prediction of human behavior, for example, pattern of life. The goal is to achieve near-perfect knowledge by increasing the rate of data collection and, therefore, understanding about the target. This enables a faster decision cycle from more detailed information. The collection goal is that a target will be unable to evade information collection. The purpose is to improve decision making while reducing risk. Unmanned aerial vehicles (UAVs) are frequently associated with persistent collection.

Open-Source GEOINT Data

Any type of lawfully and ethically collected geospatial information from publicly available sources is considered to be open-source material. Open-source is contrasted with closed-source material that is not available to the public. There is an increasing reliance on the open-source collection of information.

The cyber world can be an important source for open-source data. When viewed with an eye to intelligence analysis, social media, and more broadly, cyber transactions between individuals and groups, describe past and current events and help to anticipate future events. These transactions can be aggregated into trends, models, and conditions that describe who is involved and where and when an event will be or is happening. While much of the information disseminated in social media is not geographic information per se, such social media transactions contain massive amounts of geographic information in the "from" and "to" communication nodes and possible geographic references in the content. Significantly, such cyber activities are intrinsically self-documenting and provide spatial and temporal information to enable analysts to focus on point events, group behaviors, or larger trends. The result is that cyber activities create a vast amount of useful data about an individual or group in the context of local, regional, and global activities. Given the growing impact of technology and the information-rich content, the importance of cyber activity has and will likely continue to grow in value to the intelligence community.

Crowdsourcing GEOINT Data

Crowdsourced geospatial data (CGD) is an emerging trend that is influencing future methods for geospatial data acquisition. CGD involves the participation of untrained individuals with a high degree of interest in geospatial technology. Working collectively, these individuals collect, edit, and produce datasets. Crowdsourced geospatial data production is typically an open, lightly controlled process with few constraints, specifications, or quality assurance processes. This contrasts with the highly controlled geospatial data production practices of national mapping agencies and businesses. Adoption of CGD and production methods has been a concern, especially to government organizations, due to quality concerns related to differences in production methods.

Closed Source GEOINT Data

GEOINT data are collected openly and covertly (secretly). Data collected covertly may be considered closed-source. Some intelligence collection must remain secret as their revelation could jeopardize the individuals involved. Closed-source data are government or private data not available through open inquiry. In simple terms, it is closed-source data if the data are not meant to be openly available to the public. Closed-source data, because of its origin, are often considered more accurate and reliable.

Closed-source data include material that a government denotes "classified" in order to restrict public access so as to protect confidentiality, integrity, or availability. Closed source also includes material such as proprietary business information, law enforcement data, educational records, banking records, and medical records. In general, closed-source data are obtained or derived from sources that:

- are marked "do not make this information available to the public";
- apply protective measures to ensure the information is made available only to select persons or systems, usually marked with a security classification by the source;
- are not available on the Internet;
- are proprietary; or
- may include a copyright.

GEOINT's Tradecraft, Geographic Information Science, and Geographic Technologies

This section expands on the concept of tradecraft and focuses on the fourth GEOINT principle, that geospatial intelligence analysis tradecraft is the art of geospatial reasoning informed by *geographic information science* (GIScience) aided by *geographic information technology* (GIT). This section defines geospatial intelligence tradecraft; describes the relationship of geographic information science, geographic technologies, and GEOINT's tradecraft; and develops GEOINT using the tools and tradecraft.

The GEOINT Tradecraft

Tradecraft is how GEOINT gets done; it includes the principles, tools, and standards of rigor for analyst's thinking. GEOINT tradecraft is a merging of science and technology with the conventional notion of tradecraft. Historically, tradecraft is defined as how an intelligence officer makes clandestine contact with an agent to obtain information, how the information is passed (or processed), and the tactics for preventing detection. In a more modern sense, tradecraft is also defined as an organization's sources and methods. GEOINT sources may include information obtained clandestinely, with human or with technical collection systems (e.g., satellites), and using modern-day, open-source information (e.g., tweets). Methods pertain to the techniques used by analysts. An organization's geospatial sources and methods may be closely guarded so as not to give opponents the opportunity to know the capabilities and interests of an intelligence gathering organization. This chapter defines the GEOINT tradecraft as:

> The GEOINT tradecraft is unique and sometimes privileged organizational sources and methods for obtaining information of a place and making sense of the information to support the decision maker in understanding human activities and intentions. Methods comprise the technologic tools to organize geospatial data and the cognitive techniques used by the analyst to make sense of it when rendering judgments, insights, and forecasts.

Critically, there is an interrelationship of humans and technologies that shapes the aspects of analysis, which is why GEOINT's tradecraft has been particularly influenced by the *Geospatial Revolution*. Geospatial intelligence's analytic craft has been shaped by geographic information science and geographic technologies. The relationship might be illustrated as in Figure 6.2:

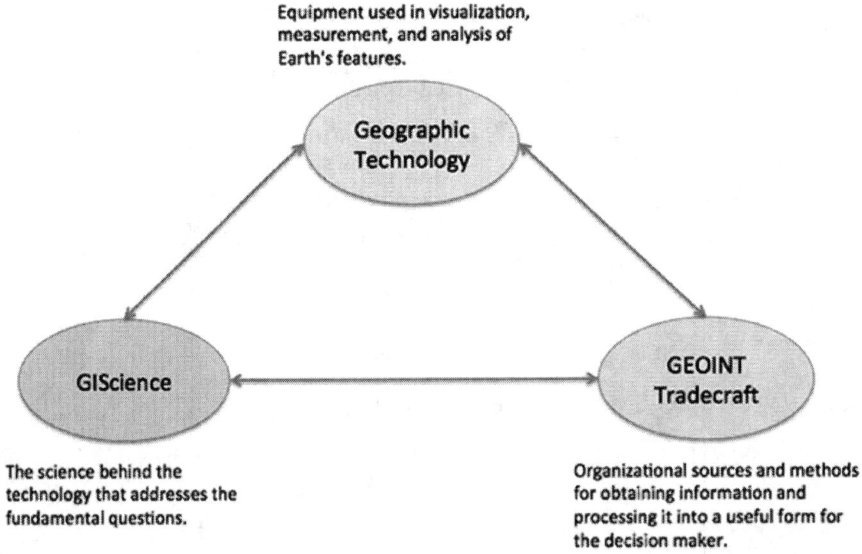

Figure 6.2. The Relationship between Geographic Technology, GIScience, and GEOINT Tradecraft.
Source: Todd S. Bacastow.

Geospatial Deception

Remote sensed imagery is a central source of GEOINT. As can be imagined, there are circumstances where people want to deny successful analysis of such imagery. This is particularly true in a military context where troops prevent enemies from performing reconnaissance by concealment, camouflage, and deception. However, it could be equally true for someone wanting to deny the ability to count the number of automobiles awaiting shipment in an automobile factory's parking lot. A core skill of the GEOINT tradecraft is contending with deceptive information. Deception should not be mistaken for the error or misinformation in a routine spatial analysis. Here, deception is designed to gain an advantage. Avoiding being deceived requires the analyst to know the ways by which concealment or obscurity is attained. This know-how includes how the method is tailored to a particular observer and how the observer might make a false judgment about the camouflaged object.

Geospatial Reasoning

Geospatial reasoning includes processes that support exploration, understanding, and sense-making. Geospatial reasoning begins with the ability to use space as a context, or framework, to make sense of what is seen. There are three spatial frames within which one can make the transition from what is observed to meaningful information; these are: behavioral spaces, physical spaces, and cognitive spaces. In all cases, the definition of the space provides an interpretive context that gives meaning to the data.

- *Behavioral space* is the four-dimensional space-time where spatial thinking is a means of coming to grips with the spatial relations between yourself and objects in the physical environment.
- *Physical space* is also built on the four-dimensional world of space-time, but focuses on a scientific understanding of the nature, structure, and function of phenomena.
- *Cognitive space* is in relationship to concepts and objects that are not in and of themselves necessarily spatial.

Learning to think spatially is to consider objects in terms of their context, or to understand the object's location in behavioral space, physical space, or cognitive space, to question why objects are located where they are, and to visualize relationships between and among these objects. The key skills of spatial thinking include having the ability to:

- understand the context;
- recognize patterns and shapes;
- recall previously observed objects;
- integrate observation-based learning; and
- mentally rotate an object and envision scenes from different viewpoints.

Sense-making

Geospatial analysis can be very difficult to do well. The difficulty is cognitive and most frequently not related to an individual's ability to use the tools that geographic information technologies provide. Good geospatial analysis requires the analyst to monitor their mental progress, make changes, and adapt the way they think. This is self-reflection, self-responsibility, and self-management. Richards Heuer makes three important points relative to intelligence analysis:[14]

- Human minds are ill equipped ("poorly wired") to cope effectively with both inherent and induced uncertainty.
- Increased knowledge of one's own inherent biases tends to be of little assistance to the analyst.
- Tools and techniques that apply higher levels of critical thinking can substantially improve analysis of complex problems.

The core of his argument is that even though every analyst sees the same piece of information, it is interpreted differently due to a variety of factors. In essence, one's perceptions are morphed by factors that are out of human control. These cognitive patterns, or mindsets, are potentially good and bad. On the positive side, they tend to simplify information for the sake of comprehension, but they also bias interpretation. The key risks of mindsets are that:

- analysts perceive what they expect to perceive;
- once formed, they are resistant to change;
- new information is assimilated, sometimes erroneously, into existing mental models; and
- conflicting information is often dismissed or ignored.

According to Heuer, mental models, mindsets, or cognitive patterns are essentially the analogous image by which people perceive information. Even though every analyst sees the same piece of information, it is interpreted differently due to a variety of factors. In essence, one's perceptions are morphed by a variety of factors that are completely out of the control of the analyst. However, cognitive patterns are critical to allowing individuals to process what otherwise would be an incomprehensible volume of information. Yet, they can cause analysts to overlook, reject, or forget important incoming or missing information that is not in accord with their assumptions and expectations. Ironically, the experienced analysts may be more susceptible to these mindset problems as a result of their expertise and past success in using time-tested mental models. Since people observe the same information with inherent and different biases, Heuer believes an effective analysis method needs a few safeguards. The analysis method should:

- encourage products that clearly show their assumptions and chains of inferences; and
- emphasize procedures that expose alternative points of view.

Heuer advocates using structuring analytical problems as a means to overcome mindsets.[15]

Structured Analytic Techniques (SAT)

Most people solve geospatial problems intuitively by trial and error. Structured analytic techniques (SAT) are a "box of tools" to help the analyst mitigate cognitive limitations and pitfalls. Structured thinking in general, and structured geospatial thinking specifically, is at variance with the way in which the human mind is in the habit of working. Taken alone, SATs do not constitute an analytic method for solving geospatial analytic problems. The most distinctive characteristic is that structured techniques help to decompose one's geospatial thinking in a manner that enables it to be reviewed, documented, and critiqued. The U.S. Government publication *A Tradecraft Primer: Structured Analytic Techniques for Improving Intelligence Analysis* highlights a few of the key structured analytic techniques used in the private sector, academia, and the intelligence profession.[16]

Structured analytic techniques help the mind think more rigorously about an analytic problem. In the geospatial realm, they ensure that key geospatial assumptions, biases, and cognitive patterns are not just assumed correct, but are well considered. The use of these techniques later helps to review the geospatial analysis and identify the cause of any error. Moreover, structured techniques provide a variety of tools to help reach a conclusion.

Analytic Stages

Structured analytic techniques and other problem solving approaches can be applied in three stages: problem initiation, information foraging, and sense-making. Table 6.3 shows how various structured analytic techniques fit with these stages.

Table 6.3. How Various GEOINT Techniques and Tools Fit Together.

Stage	Possible SAT	Geospatial Technology Operation
Stage 1: Problem Initiation	· Decomposition and visualization · Challenging mindsets · Group process techniques	· Geospatial data entry · Geospatial data conversion · Data validation · Geospatial data management · Attribute data management · Data visualization
Stage 2: Information Foraging	· Brainstorming	· Data visualization · Geospatial data processing/analysis
Stage 3: Sense-making	· Indicators, signposts, scenarios · Hypothesis generation and testing (analysis of competing hypotheses) · Challenging mindsets	· Output of maps and reports

Analytic Standards and Judgments

Professional Standards

The analytic standards of the intelligence community are the core of GEOINT's tradecraft. The following is based upon the Office of the Director of National Intelligence's *Intelligence Community Directive 203* regarding analytic standards.[17] These standards articulate a commitment to meet the highest standards of integrity and rigorous analytic thinking, and serve as goals for analysts who strive for excellence in their analytic work. The analytic standards are characterized by:

- objectivity;
- independence of political considerations;
- timeliness;

- basis on all available sources of intelligence information; and
- implementation of analytic tradecraft standards that include:

  1. Properly describes quality and reliability of underlying sources, data, and methodologies.
  2. Properly expresses and explains uncertainties associated with analytic judgments.
  3. Properly distinguishes between underlying intelligence and analysts' assumptions and judgments.
  4. Incorporates analysis of alternatives.
  5. Demonstrates customer relevance and addresses implications.
  6. Uses clear and logical argumentation.
  7. Explains change to or consistency of analytic judgments.
  8. Makes accurate judgments and assessments.
  9. Incorporates effective visual information where appropriate.

Expressing Analytic Judgments

As stated previously, intelligence is not truth. Intelligence is an approximation of truth with some level of confidence. It is necessary for decision makers to know how confident their analysts are in the results of an analysis; however, analysts are often reluctant to state the uncertainties surrounding their work because they may not:

- Appreciate the value of that information to decision makers.
- Trust the decision maker to understand any caveats as to their confidence in the work.
- Expect an honest statement of uncertainty to be rewarded.
- Know how to express the uncertainty.

U.S. Joint Publication 2–0, Doctrine for Intelligence Support to Joint Operations, suggests terms and expressions to communicate analytic judgments.[18] The terms selected are based on three factors: the number of key assumptions required, the credibility and diversity of sourcing, and the strength of argumentation. Each factor should be assessed independently and then in concert with the other factors to determine the confidence level. Multiple levels are stated as low, moderate, and high. Phrases such as "we judge" or "we assess" are used to call attention to a product's key assessment. Supporting assessments may use likelihood terms or expressions to distinguish them from assumptions or reporting.

Conclusion

Advances in technology, such as GIS, GPS, UAVs, and cellular phones, have increased the application of geospatial knowledge. This has also resulted in an expansion of the uses of geospatial intelligence. While historically applied to military and defense realms, today the principles of GEOINT extend to other fields including homeland security and law enforcement, business, and humanitarian efforts. GEOINT is the art and science of place and time on Earth's surface. Its subject matter is the human phenomena that make up the world's places. The GEOINT professional studies and describes the changing pattern of places in words, maps, and graphics, explaining how patterns come to be, unraveling their meaning, and anticipating changes in patterns for decision makers. This chapter described GEOINT data, sources, and collection methods. It examined how geographic technology, geographic information science, and the geospatial intelligence tradecraft shape the geospatial intelligence discipline and provided tools and techniques of the geospatial intelligence analyst. The principles and techniques of GEOINT are applicable to many real-world problems.

Notes

1. See https://www.cia.gov/library/center-for-the-study-of-intelligence/kent-csi/vol3no2/html/v03i2a03p_0001.htm
2. Mark M. Lowenthal, *Intelligence: From Secrets to Policy,* 6th edition (Los Angeles, CA: CQ Press, 2015), 8.
3. *U.S. Code (2006), Title 10—ARMED FORCES,* Section 467, http://www.gpo.gov/fdsys/pkg/USCODE-2006-title10/html/USCODE-2006-title10-subtitleA-partI-chap22-subchapIV-sec467.htm
4. See http://www.princeton.edu/~achaney/tmve/wiki100k/docs/HUMINT.html
5. See https://www.nsa.gov/sigint
6. Todd S. Bacastow and Dennis J. Bellafiore, "Redefining Geospatial Intelligence," *American Intelligence Journal* (Fall 2009): 40.
7. See http://www.ncgia.ucsb.edu/giscc/units/u002
8. Torsten Hägerstrand, "What about People in Regional Science?" *Papers of the Regional Science Association* 24, no. 1 (1970).
9. Harvey J. Miller, "A Measurement Theory for Time Geography," *Geographical Analysis* 37 (2005), http://onlinelibrary.wiley.com/doi/10.1111/j.1538-4632.2005.00575.x/full

10. See Gabriel Miller, "Activity-Based Intelligence Uses Metadata to Map Adversary Networks," *Defense News,* July 8, 2013, http://www.defensenews.com/article/20130708/C4ISR02/307010020/Activity-Based-Intelligence-Uses-Metadata-Map-Adversary-Networks
11. James J. Wirtz, "The Sources and Methods of Intelligence Studies," in *The Oxford Handbook of National Security Intelligence,* ed. Loch K. Johnson (New York: Oxford University Press, 2010).
12. National Geospatial-Intelligence Agency, National System for Geospatial Intelligence, *Geospatial Intelligence (GEOINT) Basic Doctrine Publication 1–0* (Office of Geospatial-Intelligence Management, September 2006), 10.
13. Information regarding the Intelligence Cycle can be found in: Commission on the Intelligence Capabilities of the United States Regarding Weapons of Mass Destruction, *Report to the President of the United States* (Washington, D.C., March 31, 2005), 583–584.
14. Richards J. Heuer, Jr., *Psychology of Intelligence Analysis* (Washington, D.C.: Center for the Study of Intelligence, 1999), xviii.
15. Ibid., 85.
16. Central Intelligence Agency, *A Tradecraft Primer: Structured Analytic Techniques for Improving Intelligence Analysis* (Center for the Study of Intelligence, March 2009).
17. Office of the Director of National Intelligence, *Intelligence Community Directive 203, Analytic Standards,* January 2, 2015.
18. Joint Chiefs of Staff, *Joint Publication 2–0, Joint Intelligence,* October 23, 2013, A-2, http://www.dtic.mil/doctrine/new_pubs/jp2_0.pdf

7

Use of "Drones" in Homeland Security: A Comparative Perspective on Use of Security Technology and Its Legal, Political, and Social Aspects

ANDREA JERKOVIĆ & ALEXANDER SIEDSCHLAG

Introduction

The increase in surveillance technologies evokes new debates between supporters and opponents in a variety of fields,[1] and has recently especially done so regarding the use of unmanned aerial vehicles, or UAV, commonly known as "drones."[2] A few years ago, use of drones would have been primarily associated with the use of, for example, Predator and Reaper-class systems for "targeted killings" in the Global War on Terror. Today a much more differentiated perception exists, also in the public, following increasing media coverage on domestic use of drones in business (such as Amazon's exploration of drone-based parcel delivery) as well as in homeland security and emergency management (including border surveillance and disaster relief support), with criticism also being more differentiated.[3] We should specifically speak about *unmanned aerial systems,* or *UAS,* since we are talking not only about a flying object but about big data—sourcing and processing of information from interconnected systems—and about payloads (detection, surveillance, and other possible technology) carried.

Future regular use of UAS in homeland security will increasingly include Earth-observation-based techniques for critical infrastructure designation, such as in the context of big events, thus supporting mainstreaming of situational awareness and common situational picture generating processes across jurisdictions and agencies.[4] For example, UAS are useful in mapping out the surroundings of a stadium to help law enforcement determine critical infrastructure and protection needs in the area.[5] They will also be used more and more in emergency management, for instance, in the analysis of disaster sites. UAS can enable timely, comprehensive documentation of critical infrastructures impacted by different types of disasters, also supporting evidence-based scenario planning for better emergency management and law enforcement responses to disasters.[6]

The European Union security research[7] project AEROCEPTOR ("UAV-based Innovative Means for Land and Sea Non-Cooperative Vehicles Stop") explores the combination of drone-based surveillance with drone-driven security enhancing measures, based on payload designed to actually interfere with the movement of non-cooperative vehicles operated by terrorists or criminals.[8] This chapter—focused on the U.S. and the EU—presents some results from a comparative legal, political, and societal framework analysis that was conducted as part of this project.

United States

Legal Framework

The U.S. Customs and Border Protection (CBP)—a component of the U.S. Department of Homeland Security—is currently the only agency permitted to use UAS on a daily basis within the nation's borders. However, use of drones in U.S. homeland security was already explored in 2003 and since then has evolved along the policy framework that was created by Congress.[9] This was accelerated by the fact that CBP is flying operations for, or lending drones to, other agencies:

> Most of the missions performed are for the Coast Guard, the Drug Enforcement Administration (DEA), and immigration authorities. Other federal, state, and local law enforcement agencies have been borrowing border patrol drones for surveillance missions, but the drones also serve in disaster relief, drug seizures, and in the search for missing persons.[10]

Foundational legislative mandates included the following, with a focus on border surveillance:

- A congressional amendment to the Fiscal Year 2003 Department of Defense Authorization Act required the president to issue a report "on the use of unmanned aerial vehicles for the support of homeland security missions."

- The Intelligence Reform and Terrorism Prevention Act of 2004 mandated that the homeland security secretary "'shall design the pilot program' that would examine the 'use of advanced technological systems, including sensors, video, and unmanned aerial vehicles, for border surveillance.'"
- House and Senate members in northern border states included language in the Intelligence Reform and Terrorism Prevention Act of 2004 requesting DHS to test the feasibility of using unmanned aircraft to patrol the northern border of the United States.[11]

This raised the issue of (surveillance) data collection by drones used in homeland security. In 2007, responding to the growing concern about big data, the Federal Aviation Administration developed a concept for "System Wide Information Management," or SWIM, with its value to helping ensure consistency and safety of UAV flight operations being discussed as well.[12] SWIM could help address, among others, aspects such as the following:[13]

- classification of data security levels;
- dynamic data exchange;
- validation of data against authoritative sources;
- aerial reconnaissance;
- crowd monitoring;
- supply chain monitoring;
- communications (e.g., telecommunication, telecom relay, signal coverage, etc.);
- survey (e.g., geographic surveys and infrastructure, etc.); and
- disaster management (e.g., rescue measures, damage estimation, etc.).

In 2012, Congress passed the FAA Modernization and Reform Act of 2012, directing the Secretary of the Department of Transportation

> to determine whether UAS operations posing the least amount of public risk and no threat to national security could safely be operated in the NAS [National Airspace System] and if so, to establish requirements for the safe operation of these systems in the NAS, prior to completion of the UAS comprehensive plan and rulemakings."[14]

Following a requirement from the Modernization and Reform Act, in February 2015 the Federal Aviation Administration (FAA) proposed a "framework of regulations that would allow routine use of certain small unmanned aircraft systems (UAS) in today's aviation system, while maintaining flexibility to accommodate future technological innovations."[15] Since 2012, when the Act was passed, "15 states have enacted laws restricting drones in some way, according to data from the National Conference of State Legislatures (NCSL)."[16] The latter mainly is a result from concerns related to the Fourth Amendment of the U.S. constitution that protects citizens from government intrusion.[17]

Court Decisions and Public Acceptance

Civil society organizations are becoming vocal in the privacy debate over UAS, and they have directed general and specific concerns to FAA. This includes potential for abuse ("drone hacking" or interception of a legal drone for illegal use); detailed data collection statements as precondition for licensing civilian UAV use; and the proliferation of UAS technology as a serious threat to privacy and other civil liberties.[18] The legal response to use of surveillance technology is evolving accordingly.[19] The Supreme Court decided that law enforcement permits the use of surveillance technology via a private plane in "public navigable airspace," including over enclosed backyards, and that this did not pose a Fourth Amendment violation.[20] When more sophisticated tracking technology is used, the Supreme Court sees this as a "search" under the Fourth Amendment, thus requiring a legal warrant.[21] However, the Supreme Court ruled, routine surveillance and searches conducted at the border do not require warrant or probable cause.[22]

Specific polls have been conducted in the U.S. to develop a picture of citizens' perceptions of and attitudes toward domestic use of drones in homeland security and elsewhere. A poll conducted by Monmouth University in 2013 showed that a majority recognizes the societal and economic benefits of UAS technology, as well as its use in homeland security and in law enforcement in general.[23] For example, 62% of the respondents supported "the use of drones to control illegal immigration on the nation's border;" 83% were supportive of "the use of drones to help with search and rescue missions;" and 52% of "allowing law enforcement to use drones armed with weapons in hostage situations." However, only 44% were in favor of "allowing law enforcement to use drones armed with weapons to patrol the nation's border," which shows that the public differentiates between uses of the same technology in different mission scenarios.

Addressing Ethical Aspects, Regulatory Limitations, Dual Use, and Risks of Abuse

The American Civil Liberties Union cautioned that UAS raise specific constitutional issues, of a different order to manned aircraft:

> Because of their potential for pervasive use in ordinary law enforcement operations and capacity for revealing far more than the naked eye, drones pose a more serious threat to privacy than do manned flights. There are good reasons to believe that they may implicate Fourth Amendment rights in ways that manned flights do not.[24]

The American Civil Liberties Union thus recommended core measures to preserve the privacy issues, including usage restrictions, image retention restrictions, public notice, democratic control, as well as auditing and effectiveness tracking.[25] Technology as such is value free and knows no intentions, whether good or bad, or civil or military.[26] The human factor aspect clearly needs to gain a more prominent role in UAS debates in the U.S. and elsewhere, in addition to addressing constitutional, legal framework, and regulatory aspects.

In early 2013, FAA proposed small UAS-related rulemaking that focuses on the following dimensions that include, among other things, addressing of the human factor as it relates to operators and covers both safety and security aspects:[27]

- "Operator Certification and Responsibilities": Pilots of small UAS would qualify as "operators," which means they would need to pass an aeronautical knowledge test; obtain a certificate with type rating; be checked by the Transportation Security Administration (TSA); provide the UAS to FAA, upon request, for inspection or testing; etc.
- "Operational Limitations": Those include safety and security-by-design principles such as weight limits ("unmanned aircraft must weigh less than 55 lbs. [25 kg]"); maximum airspeed (100 mph) and altitude (500 feet above ground level); visual-line-of-sight requirement ("at all times the small unmanned aircraft must remain close enough to the operator for the operator to be capable of seeing the aircraft with vision unaided by any device other than corrective lenses"); anti-mission creep provisions, that is, "small unmanned aircraft may not operate over any persons not directly involved in the operation"; etc.
- Extent of applicability of aircraft requirements and applicability of the proposed rulemaking to model aircraft.

The FAA approach to UAS rulemaking also represents a societal security approach since they published the proposed regulation on the Internet and public comment was invited.[28]

UAS are dual-use technologies, usable for both civil and military-sector deployment. A RAND corporation study on *The Global Technology Revolution 2020,* completed in 2006, had already stressed that dual-use concerns should not lead to an overemphasis of this risk to militarize UAS. The broad spectrum of safety and security-related uses should be considered, including for homeland security and emergency management purposes, such as monitoring and surveillance, but also support to fire fighting and response to natural disasters.[29] Other considerations have addressed the risk of adversarial abuse, where "a country or terrorist group motivated to develop a crude cruise missile or UAV either on its own or with some foreign assistance could readily take advantage of the last decade's quantum leap in dual-use technologies that comprise the chief components of autonomous air-vehicle development."[30] Broader contemplations of ethical issues related to unmanned systems have called for a designing-in approach, so that technology development would integrate anti-tamper measures into dual-use and military systems that reduce risk of adversaries gaining information.[31]

This is an important futuristic contribution to the discussion of homeland security technology, should UAS at some point in the future be operating in full autonomous mode for border surveillance, law enforcement, counterterrorism, and other purposes. Drone vulnerability to "terrorist hijacking" had already made it into the news in 2012,[32] and a Congressional Research Services report of the same year among other things summarized potential security risks.[33] It further called for a "comprehensive approach" to governance of drone-related risks, noting that while under the Aviation and Transportation Security Act, "responsibility for aviation security was transferred from FAA to the newly formed Transportation Security Administration (TSA) in 2001 […], TSA has not specifically addressed the potential security concerns arising from the operation of drones in domestic airspace."[34]

European Union

Application Sectors and Markets in Europe

Similar to the U.S., debate about use of UAS in the EU has largely been triggered by border management and surveillance issues. With the evolution of the EU's border agency FRONTEX and the EUROSUR border system,[35] it has been suggested, for instance, that European research and development programs be targeted toward improving

> the performance of surveillance tools and sensors (e.g. satellites, unmanned aerial vehicles/UAVs, etc.) and developing a common application of surveillance tools. A common pre-frontier intelligence picture could be developed to combine intelligence information with that obtained from surveillance tools.[36]

Discussion of the use of UAS in homeland security in the EU—or in "home affairs" in the EU jargon—is, however, part of a larger policy debate. It not only focuses on law enforcement and government use of UAS for surveillance purposes, but also addresses the commercial use of drones, including their potential use by marketing firms, private investigators, etc.[37] The discussion falls within broader exploration of "key civilian application markets," which a European Commission study of 2007 had identified as law enforcement, firefighting, energy sector, agriculture forestry and fisheries, Earth observation and remote sensing, as well as communications and broadcasting.[38]

Legal Framework

Implementation considerations in the EU, similar to the U.S., often focus on the issue of integrating unmanned aircraft into civilian-controlled, non-segregated airspace, with an emphasis on safety concerns.[39] The discussion has been criticized for not being detailed enough as far as targeted surveillance and data protection issues are concerned, and it has been suggested that protection of critical infrastructure should be more commonly defined as an important civilian task that can be accomplished with the help of UAS.[40] Discussion of ethical, legal, and social issues (ELSI)[41] of use of UAS in the EU typically is embedded in the larger context of the ghost of the surveillance society and how surveillance technologies in general have an impact on a wide range of societal values, freedom, and individual rights.[42] Further, discussion about the use of UAS in the EU is framed in terms of gaps between the evolution of technology, uses of technology by law enforcement and other authorities, the legal and regulatory context, as well as the evolution of counterterrorist policy.[43] Similar to the U.S., anticipatory regulatory strategies and "design-stage controls" (i.e., designed-in limitations for abuse) are called for.[44] The emphasis here is not on limiting physical abuse of a drone, but on limiting, through a "privacy-by-design" approach, the potential for abuse of the (surveillance) data it provides, and/or their unauthorized processing.[45]

This focus, to a large extent, results from the existing EU legal framework:

> All actions related to the development of the RPAS[46] must respect the rights and principles enshrined in the Charter for Fundamental Rights of the EU, and in particular the right to private life and family life (Article 7) and the protection of personal data (Article 8). [...]
>
> *Directive 95/46/EC* of the European Parliament and of the Council of 24 October 1995 on "the protection of individuals with regard to the processing of personal data and on the free movement of such data" sets out the data protection legal framework. [...] It applies also to RPAS, because RPAS do in principle not add new features to already available information collection systems such as manned aircraft, satellites or cameras.[47]

Practically, the principles for a regulatory approach are similar to that of the FAA in the U.S., whereas the implementation deadline for EU legislation is 2028 only. New EU rules to be introduced include UAS certification, safety certification, remote pilot/operator licensing, as well as data protection and privacy. Different from the U.S., EU regulations are expected to cover vehicles up to 150 kilos of weight.[48]

EU Member State Approaches, Addressing of Ethical Issues, and Public Acceptance

EU Member States have different policies for domestic use of UAS, as well as legal frameworks that lay emphasis on different aspects, for example:

- In France, UAS are already in use to monitor crowds and demonstrations for large event security purposes. The legal grounds for data processing are a major issue.[49]
- In Germany, surveillance and protection of critical infrastructure are the main aspects, and privacy and data protection are the main issues.[50]
- In Italy, UAS so far are only used in military application. Police have, however, started to explore crime prevention applications, which appears to be supported by public opinion, but there is no discussion about the use of UAS for homeland security purposes, other than in military theaters in the war against terror.[51]
- In Spain, mainly following safety and airworthiness concerns, UAS are currently only used for experimental purposes authorized by the Spanish Civil Aviation General Directorate. Media tend to portray UAS as robots

that make decisions on their own. This inaccurate picture reduces public acceptance of UAS-related technology also as it applies to civil security.[52]

- In the UK, use of UAS for policing purposes and large event security, including crowd control and monitoring of suspects, has become increasingly common since around 2006.[53] However, in the UK, UAS are required to meet the same safety and operational standards as manned aircraft, and while "there are no specific data protection regulations which apply to drones, nor any specific guidance on the subject [...], personal data collected through drones would have to comply with the Data Protection Act: a valid ground for processing data would be needed (e.g. legitimate interest of the controller), information should be given to the data subjects, and the UK Data Protection Agency would have to be notified of the data processing activity."[54]

EU Member States differ as well on approaches to ethical issues related to use of UAS in home affairs, and there is no common view on how technical requirements should be shaped to address ethical concerns. However, like in the U.S., it has been pointed out that UAS need to be assessed in a differentiated way, reaching a balanced view between (mainly privacy-related) concerns and the useful missions that this technology is able to support, such as improving civil protection and disaster response, border surveillance, as well as establishing a growth market.[55] These suggestions are supported by Eurobarometer opinion poll results, according to which "EU citizens think that robots should be used as a priority in areas that are too difficult or dangerous for humans, like space exploration (52% priority), manufacturing (50%), military and security (41%) and search and rescue tasks (41%)."[56] At the same time, EU discussions on public acceptance of UAS use in home affairs has a focus on targeted surveillance activities and sometimes still become mixed up with the criticism of "targeted killings" by drones in military use.[57]

Conclusion

We are unlikely to see a comprehensive approach to use of UAS in homeland security, or for other purposes, in and among the U.S., the EU and its Member States, and other countries. The main reason is that there are no clear legal definitions of "public space" in which the use of such new technologies should be regulated.[58] While following a similar thread, U.S. debates typically center on the question of whom to categorize as a target and on surveillance and information-gathering purposes vis-à-vis citizens' constitutional rights. EU discussions emphasize privacy by design, focusing on aspects immanent to technology that are increasingly being addressed in the U.S. as well. Reflections in both the EU and the U.S. converge to the point that as UAS technology and implementation increases, discussion of application scenarios should include technology (risk) assessment, addressing the growing autonomy and intelligence of such technology, including the problem of captured or hijacked technology.

In addition, the human factor of operator qualification and training is emphasized in the U.S. and the EU, though in the EU the situation is more complicated because training for platform operators is diversified and controlled by authorities at different levels. The U.S. is capable of pursuing a more unified approach, but it may tend to overlook the downside to a "'one-size-fits-all'" approach, which may be inappropriate given the diversity in UAS technologies and operational missions.[59] When addressing dual-use issues, there are risks as well as opportunities involved. Often, analysis and public debate focuses on a risk of "militarization" of "drones" domestically, whereas it is often overlooked that professional military education often comprehensively addresses ethical and policy discussions regarding emerging technologies (such as military robotics) and their use in the security sector. Civil-military collaboration in UAS-related training could thus be of benefit.[60]

Discussions both in the EU and the U.S. emphasize that public acceptance depends on the development of a transparent market for civil UAS applications, and awareness of their economic and other benefits. Furthermore, public acceptance is closely connected to security cultures and the perception of the risks and benefits of technology: The use of UAS for general civil security purposes will need some time to achieve general citizen acceptance, in particular with regard to the risk of data protection violations. As long as there are no comprehensive rules and regulations for the civil use of UAS, citizens' fear for loss of privacy and potential for limitless application of this surveillance technology is real and rational. Awareness needs to grow for the fact that UAS can carry different kinds of payload that do not necessarily need to be limited to surveillance technology. Payload, for example, may also include means of intervention such as paint balls or tire-puncturing devices in order to stop terrorist or criminal vehicles.[61] Further, when discussing potential for abuse of weaponized UAS, the debate needs to reach beyond information security aspects and address risks such as drone-based delivery of an improvised explosive device (IED) or bioagent.

As UAS assume their place in homeland security, their use by the military will further increase as well, and this will bring up further dual-use questions and need for public policy debate, informed by structured academic and think-tank analysis. This technology will allow the U.S. and other countries to maintain a presence in global hot spots and to disrupt Al Qaeda and other terrorist groups, as budget cuts downsize the military. While this will help increase the security of those and their citizens from terrorism, the question is at what price. Will it carry the risk of infringing constitutionally enshrined fundamental values—such as personal freedom and privacy, possibly moving the countries toward a surveillance-based society—or the risk of creating technology that puts societies as much at risk as it protects them? Research on science and technology for homeland security has pointed out that increase in technological capabilities does not automatically decrease vulnerabilities and that technology can be abused and fail, so that homeland security capabilities themselves require risk management.[62] This has been exemplified by a manned aircraft case: The intentional crashing of Germanwings flight 9525, where the mandatory post-9/11 reinforced cockpit door substantially facilitated the co-pilot's plan to kill himself and massacre all passengers by flying the plane into a mountain in the French Alps.[63] The Germanwings case also reinforces the relevance of balancing safety vs. security demands in the use of advanced technology, taking into account hardware, software, and the human factor.[64]

It is crucial to gear homeland and civil security studies to enhancing critical understanding of the benefits and risks of emerging security technologies. There is no need to reinvent the wheel for judging new surveillance, but state-of-the-art civil security research has criteria to offer that can be applied to making decisions about use of UAS domestically, and about acceptable scenarios for that new instrument of statecraft in providing security to citizens without sacrificing democracy to technology. Those include, for instance: formal procedure and public input in the decision to adopt the technology; assessing of the symbolic meaning the use of technology may imply; reversibility (the use of technology for surveillance should not permanently change state or society); human review (technology should not be automated and make decisions without the human in the loop); appropriateness and clarity of goals; proportionality (technology and information collected should be scaled to the dimension and impact of the security issue addressed); criteria of selection for the subject(s) of surveillance; as well as equal access to and periodic review of surveillance tools.[65]

Notes

The research leading to these results has received funding from the European Union Seventh Framework Programme (FP7/2007–2013) under grant agreement n° 285144 (AEROCEPTOR project, http://www.aeroceptor.eu). This publication reflects only the authors' views and the Union is not liable for any use that may be made of the information contained therein.

1. For a comprehensive state-of-the-art account, see Francesco Flammini, Roberto Setola, and Giorgio Franceschetti, eds. *Effective Surveillance for Homeland Security: Balancing Technology and Social Issues* (Boca Raton, FL: CRC Press, 2013).
2. In Europe, the term Remotely Piloted Aerial System (RPAS) is still common, however, in the U.S. this term is largely associated with military use since it is mainly U.S. Air Force terminology.
3. Tom Barry, "Drones over the Homeland. How Politics, Money and Lack of Oversight Have Sparked Drone Proliferation, and What We Can Do," *International Policy Report* (Center for International Policy, Washington, D.C.), April 2013, 12, http://www.ciponline.org/images/uploads/publications/IPR_Drones_over_Homeland_Final.pdf
4. International stakeholder consultation, conducted in summer 2013 in the project "UAV-Based Innovative Means for Land and Sea Non-Cooperative Vehicles Stop" (AEROCEPTOR), EU security research project, last accessed March 11, 2015, http://www.aeroceptor.eu
5. Ibid.
6. See "Infrastructure Protection: Drones to Help Assess Post-Disaster Infrastructure Damage," *Homeland Security News Wire,* March 6, 2015, http://www.homelandsecuritynewswire.com/dr20150306-drones-to-help-assess-postdisaster-infrastructure-damage
7. See Michel F. Bosco, "Trends on Security Research in Europe," Chapter 12 in this volume, 205–227.
8. See "UAV-Based Innovative Means for Land and Sea Non-Cooperative Vehicles Stop" (AEROCEPTOR).
9. Chad C. Haddal and Jeremiah Gertler, "Homeland Security: Unmanned Aerial Vehicles and Border Surveillance," Congressional Research Service report, RS21698, July 8, 2010, http://fas.org/sgp/crs/homesec/RS21698.pdf
10. "CBP Flew Its Drones on Behalf of Other Agencies," *Homeland Security News Wire,* January 22, 2014, http://www.homelandsecuritynewswire.com/dr20140122-cbp-flew-its-drones-on-behalf-of-other-agencies
11. Barry, "Drones over the Homeland," 12.
12. "SWIM Program Overview," Federal Aviation Administration, last published February 6, 2015, http://www.faa.gov/nextgen/programs/swim/overview

13. Matthew T. DeGarmo, *Issues Concerning Integration of Unmanned Aerial Vehicles in Civil Airspace* (McLean, VA: MITRE Center for Advanced Aviation System Development, 2004), http://www.mitre.org/sites/default/files/pdf/04_1232.pdf; Frank Sauer and Niklas Schörnig, "Killer Drones: The 'Silver Bullet' of Democratic Warfare?" *Security Dialogue* 43 (2012).
14. U.S. Department of Transportation, Federal Aviation Administration, "Operation and Certification of Small Unmanned Aircraft Systems," Billing Code 4910-13-P, RIN 2120-AJ60, February 15, 2015, 9, https://www.faa.gov/regulations_policies/rulemaking/recently_published/media/2120-AJ60_NPRM_2-15-2015_joint_signature.pdf
15. "Press Release—DOT and FAA Propose New Rules for Small Unmanned Aircraft Systems," Federal Aviation Administration, published February 15, 2015, http://www.faa.gov/news/press_releases/news_story.cfm?newsId=18295
16. Kaja Whitehouse, "Tech Crackdown on Drones Goes Local," *USA Today*, February 23, 2015, http://www.usatoday.com/story/tech/2015/02/23/crackdown-drones-technology-faa-nyc-councilman/23696377
17. Richard M. Thompson III, "Drones in Domestic Surveillance Operations: Fourth Amendment Implications and Legislative Responses," Congressional Research Service report, R42701, April 3, 2013, http://fas.org/sgp/crs/natsec/R42701.pdf
18. Louisa Brooke-Holland, "Unmanned Aerial Vehicles (Drones): An Introduction," House of Commons Library, April 25, 2013, http://www.parliament.uk/briefing-papers/SN06493.pdf
19. Thompson, "Drones in Domestic Surveillance Operations."
20. California v. Ciraolo, 476 U.S. 207, 2013 (1986); see Thompson, "Drones in Domestic Surveillance Operations," 7.
21. United States v. Jones, 132 S. Ct. 945 (2012); see Thompson, "Drones in Domestic Surveillance Operations," 9.
22. United States v. Montoya De Hernandez 473 U.S. 531, 538 (1985); see Thompson, "Drones in Domestic Surveillance Operations," 10.
23. "U.S. Supports Unarmed Domestic Drones," Monmouth University Poll, August 15, 2013, http://www.monmouth.edu/assets/0/32212254770/32212254991/32212254992/32212254994/32212254995/30064771087/409aecfb-3897-4360-8a05-03838ba69e46.pdf
24. Jay Stanley and Catherine Crump, "Protecting Privacy from Aerial Surveillance: Recommendations for Government Use of Drone Aircraft," American Civil Liberties Union (ACLU), report, December 2011, 14, http://www.aclu.org/files/assets/protectingprivacyfromaerialsurveillance.pdf
25. Ibid., 15–16.
26. Cf. Jocelyn Mawdsley, "A European Agenda for Security Technology: From Innovation Policy to Export Controls," Flemish Peace Institute, report, January 2013,

http://www.statewatch.org/news/2013/feb/fpi-european-agenda-security-technology.pdf

27. Federal Aviation Administration, "Overview of Small UAS Notice of Proposed Rulemaking," last published March 18, 2015, https://www.faa.gov/uas/nprm. See also Department of Transportation, Federal Aviation Administration, "Operation and Certification of Small Unmanned Aircraft Systems."
28. "Small UAS Notice of Proposed Rulemaking (NPRM)," Federal Aviation Administration, last published February 23, 2015, https://www.faa.gov/uas/nprm
29. Richard Silberglitt et al., *The Global Technology Revolution 2020. In-Depth Analyses. Bio/Nano/Materials/Information Trends, Drivers, Barriers, and Social Implications* (Santa Monica, CA: RAND, 2006), 23 and 217, http://www.rand.org/content/dam/rand/pubs/technical_reports/2006/RAND_TR303.pdf
30. Dennis M. Gormley and Richard Speier, "Controlling Unmanned Air Vehicles: New Challenges," *Nonproliferation Review* 10, no. 2 (2013): 2, http://cns.miis.edu/npr/pdfs/102gorm.pdf
31. Robert Sparrow, "Building a Better War Bot: Ethical Issues in the Design of Unmanned Systems for Military Applications," *Science and Engineering Ethics* 15, no. 2 (2009).
32. John Roberts, "EXCLUSIVE: Drones Vulnerable to Terrorist Hijacking, Researchers Say," *Fox News,* June 25, 2012, http://www.foxnews.com/tech/2012/06/25/drones-vulnerable-to-terrorist-hijacking-researchers-say
33. Elias Bart, "Pilotless Drones: Background and Considerations for Congress Regarding Unmanned Aircraft Operations in the National Airspace System," Congressional Research Service report, R42718, September 10, 2012, 12–14, http://www.fas.org/sgp/crs/natsec/R42718.pdf
34. Ibid., 14.
35. See Michel F. Bosco, "Trends on Security Research in Europe," Chapter 12 in this volume, 210–213.
36. Commission of the European Communities, "Examining the Creation of a European Border Surveillance System (EUROSUR)," COM(2008) 68 final, Brussels, February 13, 2008, 5, http://eur-lex.europa.eu/LexUriServ/LexUriServ.do?uri=COM:2008:0068:FIN:EN:PDF
37. Cf. Mawdsley, "A European Agenda for Security Technology."
38. European Commission, Enterprise and Industry Directorate-General, "Study Analysing the Current Activities in the Field of UAV. Second Element: Way Forward," ENTR/2007/065, 7, http://ec.europa.eu/enterprise/policies/security/files/uav_study_element_2_en.pdf
39. Ibid., 87; Ministry of Defence, "Joint Doctrine Note 2/11: The UK Approach to Unmanned Aircraft Systems," March 30, 2011, https://www.gov.uk/government/uploads/system/uploads/attachment_data/file/33711/20110505JDN_211_UAS_v2U.pdf

40. Therese Skrzypietz, "Unmanned Aircraft Systems for Civilian Missions," *BIGS Policy Paper* (Brandenburg Institute for Society and Security), no. 1/February 2012, http://www.bigs-potsdam.org/images/Policy%20Paper/PolicyPaper-No.1_Civil-Use-of-UAS_Druckversion.pdf
41. See Katerina Hadjimatheou, Tom Sorell, and John Guelke, "Ethics, Legal, and Social Issues (ELSI) in Homeland and Civil Security Research and the European Union Approach," Chapter 10 in this volume, 177–194.
42. "Increasing Resilience in Surveillance Societies" (IRISS), EU security research project, deliverable 1.1, "Surveillance, Fighting Crime and Violence," February 1, 2012, http://irissproject.eu/wp-content/uploads/2012/02/IRISS_D1_MASTER_DOCUMENT_17Dec20121.pdf
43. Ibid.
44. Ibid., 16 and 301.
45. Alfredo Roma, "European Commission Unmanned Aircraft System Panel Process: 4th Workshop on Societal Dimension," November 16, 2011, http://ec.europa.eu/enterprise/docs/uas/Report_Workshop_4.pdf
46. See note 2.
47. Council of the European Union, "Towards a European Strategy for the Development of Civil Applications of Remotely Piloted Aircraft Systems (RPAS)," 13438/12, Brussels, September 6, 2012, 21–22, http://register.consilium.europa.eu/pdf/en/12/st13/st13438.en12.pdf
48. Nikolaj Nielsen, "Ashton Calls for Military-grade Drones in EU Airspace," *EUobserver,* October 22, 2013, http://euobserver.com/defence/121854
49. AEROCEPTOR project, international stakeholder consultation, conducted in summer 2013; Gabriel Voisin, "Drones: Privacy Implications across the EU," Bird & Bird, July 15, 2013, http://www.twobirds.com/en/news/articles/2013/global/drones-privacy-implications-across-the-eu
50. AEROCEPTOR project, international stakeholder consultation, conducted in summer 2013; Voisin, "Drones: Privacy Implications across the EU."
51. AEROCEPTOR project, international stakeholder consultation, conducted in summer 2013; Voisin, "Drones: Privacy Implications across the EU."
52. AEROCEPTOR project, international stakeholder consultation, conducted in summer 2013; Voisin, "Drones: Privacy Implications across the EU."
53. John A. Wallace, "Integrating Unmanned Aircraft Systems into Modern Policing in an Urban Environment" (Thesis, Naval Postgraduate School, Monterey, CA, 2012), 26, http://calhoun.nps.edu/public/bitstream/handle/10945/17474/12Sep_Wallace_John.pdf?sequence=1
54. Voisin, "Drones: Privacy Implications across the EU."
55. Roma, "European Commission Unmanned Aircraft System Panel Process."

56. European Commission, "Special Eurobarometer 382: Public Attitudes towards Robots. Summary," September 2012, http://ec.europa.eu/public_opinion/archives/ebs/ebs_382_sum_en.pdf
57. See Brian Beary, "The Drone Debate Growths in Europe," The European Institute, November 2012, http://www.europeaninstitute.org/index.php/162-european-affairs/ea-november-2012/1651-the-drone-debate-grows-in-europe
58. Discussion at the Drones & Aerial Robotics Conference (DARC), New York University (NYU) School of Law, October 11–13, 2013, last accessed March 19, 2015, http://droneconference.org
59. Elias, "Pilotless Drones," 14.
60. DeGarmo, *Issues Concerning Integration of Unmanned Aerial Vehicles in Civil Airspace.*
61. See the EU security research project AEROCEPTOR, http://www.aeroceptor.eu, last accessed March 18, 2015.
62. Jane A. Bullock, George D. Haddow, and Damon P. Coppola, *Introduction to Homeland Security: Principles of All-Hazards Risk Management,* 4th ed. (Amsterdam: Elsevier, 2013), chapter 12: "Science and Technology," 573–613.
63. "Germanwings Flight 9525," *Wikipedia,* last published March 28, 2015, http://en.wikipedia.org/wiki/Germanwings_Flight_9525
64. Yijun Yu, "Germanwings Flight 4U9525: A Victim of the Deadlock between Safety and Security Demands," *Homeland Security News Wire,* March 30, 2015, http://www.homelandsecuritynewswire.com/dr20150330-germanwings-flight-4u9525-a-victim-of-the-deadlock-between-safety-and-security-demands?page=0,0
65. Gary T. Marx, "An Ethics for the New (and Old) Surveillance," in *Effective Surveillance for Homeland Security: Balancing Technology and Social Issues,* ed. Francesco Flammini, Roberto Setola, and Giorgio Franceschetti (Boca Raton, FL: CRC Press, 2013).

8

Risk Management within Tribally Inclusive Geographic Areas: A Primer for All Risk Managers

LLOYD MITCHELL

Introduction

In homeland security governance, including emergency management, the need to follow a risk management approach and to address all tiers—federal and SLTT (state, local, territorial, and tribal)—is often emphasized, leaving it sometimes unclear what that actually means. This chapter addresses the risk management within Tribally Inclusive Geographic Areas (TIGA) as it relates to homeland security. It does so by focusing on the practical consequences of terminology and describes why understanding of a series of relevant terms can assist in risk management efforts and reduce impact of disasters or terror events at TIGA level. This is an important perspective in addition to that of resources specifically made available to tribes, such as listed in the Department of Homeland Security's *Tribal Resources Guide,*[1] and FEMA's Tribal Homeland Security Grant Program (THSGP).[2]

Not having a good working understanding of what a tribal community is can have negative consequences ranging anywhere from loss of funding due to rejection of grant applications to loss of life due to lack of communication or jurisdictional disputes in disaster responses. At the same time, it is important to understand that the basic concepts presented here that are related to working together within Tribally Inclusive Geographic Areas are not exclusive to tribal communities and thus can, and should, be considered as applicable for use throughout the United States and internationally. This chapter closes with examples of how the "working together" concepts presented herein can be applied internationally.

Finally, the reader should have a clear understanding that the discussions presented herein are not to be considered as representative of any one specific tribal community nor as tribal communities as a whole. Just like other communities, states, and nations, each tribal community is unique with its own history, culture, and future that is individual, shared with other communities, and impacted by worldwide events—both human induced and natural.

Defining Tribal Communities

This section provides a variety of viewpoints related to describing tribal communities. Some readers will agree with all or most of the descriptions while others may not agree with any. The reader's level of agreement will reflect the understandings and experiences of the reader. Some readers will accept new descriptions or create modified descriptions by incorporating new knowledge into their existing descriptions, while others will hold fast to their own understandings despite an inherent risk of bias.

Instead of trying to define tribal communities, which actually would require combining a myriad of definitions of "tribal" and "community," reviewing and analyzing them from a lexicological perspective, and presenting the final definition to an audience, would actually promote a biased approach. Rather, it is more appropriate to discuss the concepts, understandings, and acceptances of tribal communities instead of trying to promote a one-size-fits-all definition of tribal communities.

There are varieties of terms used to describe "tribal" and this is the first dilemma. Accepted definitions range from those based on innate human behavior to those of total human construct. Results include, but are not limited to, families, groups, indigenous groups, recognized tribes, unrecognized tribes, first peoples, aboriginal groups, native inhabitants, reservation inhabitants, Indians, American Indians, tribe, nation, sovereign nation, Indian country, confederated groups, bands, rancheros, allotment recipients, native corporations, and trust land residents. All of these are examples of terms related to tribal communities.

The second dilemma concerns the term community. A community, as it is widely accepted in its sociobiological sense, refers to groups of living organisms, including humans. However, it is professionally understood and widely accepted throughout political and governmental realms that human communities are further organized and constructed based upon a wide variety of factors, including but not limited to political, financial, cultural, legal, behavioral, artistic, religious, etc.

These two definition-based dilemmas create a third, definitional dilemma that is critical for risk management in Tribally Inclusive Geographic Areas: The definition of tribal communities transcends, intersects, dissects, overlaps, separates, intersperses, and brings together people, resources, and land. To make things even more complex for risk managers, the applications of the definitions of tribal communities may be based on current or historical acceptances, some of which are known primarily only within tribal communities. To increase this complexity, the definitions of tribal communities may apply permanently or temporarily, as well as semi-inclusively, partially, or even sporadically, based upon legal, cultural, ideological, etc., human constructs. Finally, to put a fourth layer of complexity into the mix for risk managers working with tribal communities, it must be understood and accepted that the application of a "tribal community" definition may also be specific for the type of disaster or terror event, funding sources, or locations.

To assist the risk management professional, an abbreviated summary table is provided below (Table 8.1) that:

- briefly describes the tribally community-related terminology, citing select examples; and
- lists for each example how risk managers working within this framework of terminology may (a) improve the effects of their measures and (b) reduce impact of disasters or terror events.

Table 8.1. Working within Tribally Inclusive Geographic Areas: Examples of Terminology for Risk Managers.[3]

Term	Brief Description, Generalized
Aboriginal Groups	Original people of a land area and their descendants; may or may not be officially recognized by a national, state, or provincial government entity. *Why understanding this term can (a) assist risk management efforts and (b) reduce impact of disasters or terror events:* 1. Aboriginal groups can be included in disaster management efforts. 2. Aboriginal knowledge and stories may have historical disaster event information.

Term	Brief Description, Generalized
	3. Aboriginal traditions and ways may provide insight into disaster management. 4. Aboriginal leaders can serve on or lead disaster and risk management teams.
Aboriginal Knowledge	Understandings of original people of a land area and the passing down of that information to their descendants. *Why understanding this term can (a) assist risk management efforts and (b) reduce impact of disasters or terror events:* 1. Aboriginal historical ecological knowledge can complement risk management. 2. Aboriginal knowledge and stories may have historical disaster event information. 3. Aboriginal traditions and ways may provide insight into disaster management. 4. Disasters may not be recognized as such, but rather as part of a natural flow. 5. Attitudes toward disaster may be indifferent or nonexistent. 6. Disaster management may take the form of acceptance rather than control.
American Indian	Indigenous group of American citizens. An individual who self identifies herself or himself with a current or historical tribal community. The official U.S. Government term for Americans with a proven lineage arising from indigenous people who inhabited areas, wholly or partially, within the United States prior to, during, or after establishment of the original, or most current, geographic boundaries. May or may not be officially recognized by a national, state, or provincial government entity. *Why understanding this term can (a) assist risk management efforts and (b) reduce impact of disasters or terror events:* 1. An American Indian may or may not be affiliated with a tribal+ community. 2. An American Indian may or may not be eligible for certain tribal services. 3. Non-specific use of the term may inadvertently include or exclude individuals. 4. It is not recommended to use this term to refer to all tribal+ communities.

Term	Brief Description, Generalized
Blood Quantum	Most recognized tribal communities in the United States require, per the U.S. Government, that in order to receive federal individual assistance outside of the reservation, college education support for example, an enrolled member be at least 25% blood quantum—meaning that he or she must have at least 25% of their blood from a member of that tribal community. Some tribes enroll members with a lesser percent. The lineage for the blood quantum is verified through birth and other tribal records and the actual blood quantum percent is counted and determined by the tribe through the tribal enrollment office, which verifies and files all documentation. *Why understanding this term can (a) assist risk management efforts and (b) reduce impact of disasters or terror events:* 1. Individuals may have differing access to services based on blood quantum. 2. Individuals may be refused certain services based on blood quantum. 3. Enrollment cards stating a person's blood quantum may be lost in disasters. 4. Disaster funds may be awarded based on number of blood-quantum qualified.
Confederated Group	A group of tribal communities, usually, but not always, small, that banded together by choice, or were put together not by choice by the U.S. Government. May or may not be officially recognized by a national, state, or provincial government entity. *Why understanding this term can (a) assist risk management efforts and (b) reduce impact of disasters or terror events:* 1. Discussions, resource sharing, and agreements can cross boundaries. 2. Confederations can act on behalf of their members, American citizens, or guests. 3. Confederations can act as a state and be given lead in disaster management. 4. Confederations can be designated as lead by FEMA in disaster mitigation.

Term	Brief Description, Generalized
	5. Risk management on band land may be the responsibility of various parties. 6. A confederated group can serve a critical role in TIGA disaster management.
Enrollment	Formal acknowledgment by a government agency or tribe or native group of an individual that she or he is either (a) an official bona fide descendent of, or (b) has the minimum establish percent of "Indian blood," also known as "blood quantum," of a recognized or an established indigenous community. Enrollment may also be at the discretion of a tribe and can include an individual who: (a) has maintained continued cultural ties; (b) has maintained a tribal residency; (c) has served the tribe heroically; or (d) other. Enrollment, depending upon status or level or type of enrollment, may allow access to resources including education, health care, and housing. In nearly all cases, records are traced back to the time of the tribal community's recognition and the first rolls, called the base roll from which current enrollees descend. *Why understanding this term can (a) assist risk management efforts and (b) reduce impact of disasters or terror events:* 1. Individuals may have differing access to services based on enrollment status. 2. Individuals may be refused certain services based on enrollment status. 3. Enrollment cards stating a person's services eligibility may be lost in disasters. 4. Disaster funds may be awarded based on number of qualified enrollees.
First Americans	The first indigenous human population groups in what is now the Americas. Depending on the geographic locations and tribal teachings, First Americans may be of Asian, Polynesian, European, Circum-Caribbean or Circum-Polar, African, or other worlds, including those unidentified and sacred. *Why understanding this term can (a) assist risk management efforts and (b) reduce impact of disasters or terror events:* 1. Historic First American sites designated by U.S. Government are protected.

Term	Brief Description, Generalized
	2. Disaster plans must incorporate First American sites designated as historic. 3. First American sites may be on, near, or far away from current tribal location. 4. There may be no currently affiliated tribe with connection to historic site. 5. Historic First American sites are required to be part of TIGA disaster plans. 6. Specialists in First American sites must be used for disaster mitigation. 7. First American sites must be protected from post-disaster events. 8. Artifacts of First Americans are to be included in disaster management plans.
Indian	Indigenous Peoples. A preferred or acceptable term for most tribal communities and people of indigenous descent in what is now the continental United States. Not a preferred term for Alaska Natives, Native Hawaiians, or Canadian First Nation residents. *Why understanding this term can (a) assist risk management efforts and (b) reduce impact of disasters or terror events:* 1. An Indian person may or may not be tribal+ community affiliated. 2. An Indian may or may not be eligible for certain tribal+ services. 3. Non-specific use of the term may inadvertently include or exclude individuals. 4. It is not recommended to use this term to refer to all tribal+ communities.
Indigenous	Having roots or historical connections to a specific geographic location or place. Some countries such as the U.S. and Canada choose to provide protections or special rights for peoples classified as indigenous. Can be legally described, via extrapolation, as "Digging In" and "Sprouting Forth," both of which correlate with the historical and agricultural Latin-based term, which is an important way to determine if a tribal group has historic ties to the land.

Term	Brief Description, Generalized
	Why understanding this term can (a) assist risk management efforts and (b) reduce impact of disasters or terror events: 1. Land base ties can be included in TIGA disaster management plans. 2. Indigenous peoples can take part in disaster management plan development. 3. Indigenous knowledge can be valuable; see "Aboriginal Groups" and "Aboriginal Knowledge."
Nation to Nation	A policy created by the U.S. Government that requires the U.S. Government to allow officially recognized tribes, bands, Rancherias, pueblos, Alaska Native Villages, etc.—as a nation in the administration of certain services—use of certain funds, and activities with foreign countries. Allows, in some activities and with varying degrees federal oversight, recognized tribal communities to: (a) function jurisdictionally as a nation when relating to foreign countries; (b) represent the United States; or (c) act in Nation to Nation capacity with foreign countries, must by recognized by the U.S. Government as having the ability to use its national status in a nation to nation capacity in a way that does not bring harm to the tribal community or the United States. See also "Treatment As A State." *Why understanding this term can (a) assist risk management efforts and (b) reduce impact of disasters or terror events:* 1. Disasters and terror events can, and do, cross boundaries. 2. Discussions, resource sharing, and agreements can cross boundaries. 3. Nations can act on behalf of their citizens and guests. 4. A nation can take the lead in disaster management.
Native American	Born in America; Born in the United States; American Indian. A more recently used term that refers to all indigenous peoples or tribal communities throughout North America. Preferred term by some individuals, but not all. *Why understanding this term can (a) assist risk management efforts and (b) reduce impact of disasters or terror events:*

Term	Brief Description, Generalized
	1. A Native American person may or may not be tribal community affiliated. 2. A Native American may or may not be eligible for certain tribal services. 3. Non-specific use of the term may inadvertently include or exclude individuals. 4. It is not recommended to use this term to refer to all tribal+ communities.
Recognition	A term used to describe whether a governmental entity officially recognizes a tribal community, and in doing so, gives special legal, financial, political, and geographical allowances, assistances, and benefits, as well as critical legal and ethical responsibilities, to the tribe, some of which tribal leaders and risk managers have to accept, allow, promote, assure, and enforce, whether they agree with them or not. *Why understanding this term can (a) assist risk management efforts and (b) reduce impact of disasters or terror events:* 1. Disasters and terror events can, and do, cross boundaries. 2. Discussions, resource sharing, and agreements can cross boundaries. 3. Recognized tribes+ must act on behalf of their members, American citizens, and guests. 4. The type or level of recognition will determine funding sources available. 5. Tribes+ can act on behalf of their members, American citizens, and guests. 6. A recognized tribe+ can act as a state, serving as lead in regional disaster management. 7. A recognized tribe+ can be designated as lead by FEMA in TIGA disaster mitigation. 8. Risk management within TIGA may be the responsibility of various parties. 9. A recognized tribe+ can serve a critical role in TIGA disaster management.

Term	Brief Description, Generalized
Reservation	Land reserved by a government for use by indigenous or native inhabitants. Term commonly used in the United States. *Why understanding this term can (a) assist risk management efforts and (b) reduce impact of disasters or terror events:* 1. Disasters and terror events can, and do, cross boundaries. 2. Discussions, resource sharing, and agreements can cross boundaries. 3. Reservations are critical components of TIGA risk management plans.
Reservation Residents	Inhabitants of lands within reservation (land reserved by a government for use by indigenous or native inhabitants) boundaries or jurisdictions. May or may not be members of indigenous populations. *Why understanding this term can (a) assist risk management efforts and (b) reduce impact of disasters or terror events:* 1. All residents are impacted by disasters in their homeland to varying degrees. 2. Tribal+ community members may have access to different or more resources. 3. Tribal+ community members may have both tribal and non-tribal residencies. 4. Non-tribal community members living in their birth land are native residents. 5. Non-tribal members living on tribal+ land cannot be denied disaster relief. 6. Non-tribal members living on tribal+ land may have limited access to resources. 7. Using "Reservation Residents" to refer only to tribal+ members not suggested.
Sovereignty	Sustaining communities without influence from outside entities. *Why understanding this term can (a) assist risk management efforts and (b) reduce impact of disasters or terror events:* 1. Accepting a nation's sovereign status may endanger risk management efforts. 2. Sovereignty, once established, can be abused.

Term	Brief Description, Generalized
	3. Sovereignty is a human construct and therefore has potential for bias. 4. Sovereign status is dynamic. 5. Risk managers must identify when sovereign status impedes risk reduction. 6. Risk managers may have to "override" sovereign status in a high-risk situation.
Sovereign Nation	An inhabited land base in which internal and external affairs necessary for sustainability are self-directed in accordance with policy established by, or accepted by, residents. May, or may not be, officially recognized by a national, state, or provincial government entity. *Why understanding this term can (a) assist risk management efforts and (b) reduce impact of disasters or terror events:* 1. Disasters and terror events can, and do, cross boundaries. 2. Discussions, resource sharing, and agreements can cross boundaries. 3. Sovereign nations can act on behalf of their citizenry and guests.
Treatment As A State	A policy created by the U.S. Government that (a) requires the U.S. Government to treat tribes+ equally as a state in the administration of certain services and the distribution of certain funds and (b) requires all U.S. states to recognize and treat tribes equally as a state in the administration of certain services and the distribution of certain funds. *Why understanding this term can (a) assist risk management efforts and (b) reduce impact of disasters or terror events:* 1. Disasters and terror events can, and do, cross boundaries. 2. Discussions, resource sharing, and agreements can cross boundaries. 3. States can act on behalf of their residents, American citizens, and guests. 4. A tribe+ can act as a state and be given the lead in disaster management. 5. A tribe+ can be designated as lead by FEMA in disaster mitigation.

Term	Brief Description, Generalized
Tribally Inclusive Geographic Area	In its most simple form, a geographic area that includes a tribal community, whether recognized as a formal entity or not by a governmental entity. *Why understanding this term can (a) assist risk management efforts and (b) reduce impact of disasters or terror events:* 1. Disasters and terror events can, and do, cross boundaries. 2. Discussions, resource sharing, and agreements can cross boundaries. 3. Tribes+ can act on behalf of their members, American citizens, and guests. 4. The type or level of recognition will determine funding sources available. 5. A tribe+ can act as a state and be given the lead in regional disaster management. 6. A tribe+ can be designated as lead by FEMA in TIGA disaster mitigation. 7. Risk management within TIGA may be the responsibility of various parties. 8. A tribe+ can serve a critical role in TIGA disaster management.
Tribe	A group of people that have come together to share interests or resources or mutual protection. A formalized group of indigenous peoples. May or may not be officially recognized by a national, state, or provincial government entity. *Why understanding this term can (a) assist risk management efforts and (b) reduce impact of disasters or terror events:* 1. Disasters and terror events can, and do, cross boundaries. 2. Discussions, resource sharing, and agreements can cross boundaries. 3. Tribes+ can act on behalf of their members, American citizens, and guests. 4. A tribe+ can act as a state and be given the lead in disaster management. 5. A tribe+ can be designated as lead by FEMA in TIGA disaster mitigation. 6. Risk management on tribal land may be the responsibility of various parties. 7. A tribe+ can serve a critical role in TIGA disaster management.

Risk Management in Tribally Inclusive Geographic Areas

This section will briefly explore risk management in Tribally Inclusive Geographic Areas, referred to with the acronym TIGA. Again, keep in mind that the term "tribally" is used inclusively to refer to the myriad of tribal communities in North America as well as Hawaii, Puerto Rico, Guam, American Samoa, and—to varying extents—northern Mexico. To begin, risk management in TIGA, as within any other type of geographic region in the world, must integrate a variety of understandings, cultures, geographies, histories, and geopolitical organizations, all the while keeping in mind fiduciary management and supporting government entities' outcome expectations, some which may be complimentary whereas others may be contradictive or even competitive.

Culturally, risk management in TIGA is really no different than any other geographic region. Respect for cultural ways and understandings must be a priority, even when dealing with urgent decisions in the midst of disasters or responses. It must be accepted and remembered that what one culture or group of people view as a disaster another culture or group of people may view as a nothing more than a natural cycle that is restoring a balance. Understandings of the underpinnings, or causes, of disasters or terror events cover a wide realm of historically, objectively, and subjectively based understandings and knowledge. Because the United States and Canadian governments, and—to varying extents—the Mexican, Pacific Island Nations, and Caribbean governments, have promoted both indigenous rights and responsibilities, a myriad of historically and culturally based administrative and fiduciary protocols have evolved, some of which may not be totally understood by non-tribal community members.

Legally, these historical and cultural protocols are often embedded into the tribal framework of laws, policies, and administrative protocols and approaches related to disaster or terror event management. Two of the most important legal constructs that risk managers must keep in mind when working with tribal+ communities within U.S. jurisdiction are "Nation to Nation" and "Treatment As A State," as explained in Table 8.1 above, citing select examples. A tribal+ community may have been legally "recognized" by the U.S. Government as a bona fide American Indian tribe, Alaska Native Village, Native Hawaiian Community, Pacific Island Community, or Pacific Island Nation. These are legal definitions that determine the type, amount, and extent of resources available to assist in risk management, including lead entities for disaster related activities from disaster planning through post disaster mitigation. Additionally, once officially, legally recognized by a government entity, a tribal community is given special legal, financial, political, and geographical allowances, assistances, and benefits, as well as critical legal and ethical responsibilities, most of which tribal+ leaders and risk managers have to accept, allow, promote, assure, and enforce, *whether they agree with them or not.*

Although this list may and will vary according to geographical and cultural factors, some of the most important things that a risk manager working with a Tribally Inclusive Geographic Area, regardless of whether she or he is a tribal member or not, should keep in mind are:[4]

- Disasters and terror events can, and do, cross boundaries.
- Discussions, resource sharing, and agreements can cross boundaries.
- Recognized tribes+ must act on behalf of their members, American citizens, and guests.
- The type or level of recognition will determine funding sources available.
- Tribes+ can act on behalf of their members, American citizens, and guests.
- A recognized tribe+ can act as a state and serve as lead in regional disaster management.
- A recognized tribe+ can be designated as lead by FEMA in TIGA disaster mitigation.
- Risk management within TIGA may be the responsibility of various parties.
- A recognized tribe+ can serve a critical role in TIGA disaster management.
- A risk manager working within a TIGA must strive toward non-discrimination.
- All TIGA risk managers are compelled to collaborate with neighboring jurisdictions.
- All residents with TIGA, whether tribal+ members or not, are to be assured services.

The bottom line is, as stated above, *all* residents with TIGA, whether tribal+ members or not, are to be assured access to, and receipt of, risk management services and the individual ethically and ultimately responsible for assuring this is the risk manager.

Extrapolating Concepts Internationally

Although this chapter has dealt specifically with tribal+ communities within Canada, Alaska, Hawaii, and the continental United States as well as Puerto Rico, Guam, American Samoa, and to varying extents, northern Mexico, the concepts presented herein can be easily extrapolated to geographic areas around the world. Working with aboriginal populations in Australia, different tribal communities in Africa, a variety of Island Nations in the Pacific, indigenous populations in South America, or tribal groups in Eastern Asia, all have similar challenges.

Risk managers must be able to, or at least strive to, work across boundaries because, as is well known, natural disasters are usually not bound by human determined national borders. With the advent of the Internet, risk managers should be able to share and access plans and resources and easily determine service gaps, keeping in mind that some of the gaps may be cultural in nature. Once various understandings are identified, risk managers can prioritize efforts and seek support from national and international entities for activity implementation.

Notes

1. U.S. Department of Homeland Security, *Tribal Resource Guide,* last published June 2013, http://www.dhs.gov/xlibrary/assets/iga/dhs-tribal-resource-guide.pdf
2. For example, "FY 2014 Tribal Homeland Security Grant Program (THSGP)," Federal Emergency Management Agency (FEMA), last published July 25, 2014, https://www.fema.gov/fy-2014-tribal-homeland-security-grant-program-thsgp
3. Terminology note: The author has taken steps to use all terms related to indigenous populations in an objective and respectful manner. As examples:
    - To encompass all terms for indigenous peoples, the term tribe or tribal with a + sign (*tribe*+, *tribal*+), will be used.
    - For indigenous peoples that have come together for a period of time and set "roots" in a geographic area, the term "tribal communities" will often be used.
    - Occasionally, the term tribe by itself will be used primarily to reference tribal communities within the continental United States, but exclusive of Native Hawaiian and Alaska Native groups.
4. See explanation in note 3.

Part III

Transversal Aspects

9

The Leadership LEAP: A New Approach for Homeland Security

R. JEFFERY MAXFIELD, RODGER BROOMÉ, & JOHN R. FISHER

Almost every significant breakthrough is a result of a courageous break with traditional ways of thinking.[1]

—STEPHEN R. COVEY

Introduction

The world is in chaos and, unfortunately, this most likely will continue. Recent news reports are replete with stories of crises. These events demonstrate the implicit and explicit need for leadership. We believe these crises and their accompanying uncertainty are the reasons people have sought strong leaders—to bring order and sanity to an otherwise uncontrollable world. This is particularly true in homeland security and the emergency services.

Crises leadership is something that has been lightly explored, but usually in the context of business or political events. Leadership in the context we present is central to human and community existence. Robin Kielkowski states:

> Leadership during crisis is essential. It is often believed to be the sole responsibility of an individual who takes command and has every answer. However, the reality is often quite different. Successful crisis response is the result of proven leadership developed through an ongoing team effort in planning long before an emergency occurs.[2]

This idea became very evident through our nation's experience with the events of September 11, 2001. We looked to members of law enforcement, fire, and emergency medical services, and later, the federal government to step into a chaotic situation and lead us out of that chaos. In fact, we as a society have venerated these services. This has also affected the emergency services' communities by creating uncertainty and confusion as to how to best prepare for and respond to incidents; hence, we have become involved in too many areas of response without the matching resources. This creates issues of safety not only for the responders, but also for the citizens they are to protect. However, the reality is that while these public servants put their lives on the line and respond to dangerous situations, they are really only managing events. We contend that along with this response, it is the actions taken on either side of a chaotic situation—the preparation, mitigation, and recovery—where true leadership is displayed. However, the collective morale of the country appears to have little regard or confidence in our present leaders.

Dr. Barbara Kellerman (2013), in her book *The End of Leadership,* contends that,

> while the leadership industry has been thriving—growing and prospering beyond anyone's early imaginings—leaders by and large are performing poorly, worse in many ways than before, miserably disappointing in any case to those among us who once believed the experts held the keys to the kingdom.[3]

Her premise is that because of technology such as the Internet, cell phones, social networking, etc., followers are finding ways to make traditional leaders, using traditional methods, less effective. In some ways, we agree with this phenomenon. However, we also see something more serious and alarming with regard to contemporary leadership. The question is, "Why are our leaders so ineffective?" We feel there may be a simple explanation. If our premise is correct, it will take much effort from scholars, those in positions of authority, and most importantly, parents and teachers to educate and implement a shift in thinking.

Our modern world operates largely on the background of scientific and technological advancement. Martin C. Dillon points out that the underlying goal of science evolved from seeking truth to seeking certainty.[4] Many people do not see these goals as different things. However, seeking certainty through science means that all accepted knowledge that is posited as "scientific" is based on empirical proof, or in other words, something that can be seen or based on physical evidence and mathematically measured.

The principal founder of behavioral science, James Watson, declared the scientific study of psychology (and by extension, sociology and other social sciences) as an empirical science based on behavior and not consciousness.[5] In other words, psychology was no longer research on the mind directly but a study of human behavior that would provide "the knowledgeable" with the ability to predict and control human behavior. From this point, psychology and the social sciences have based their values on the principles of the natural sciences, thereby regarding human beings as subjects or things under study, opening up a horizon for a whole new "means-to-ends" project.

However, leadership is holistic, not prescriptive. This may be where people have gone wrong in their thinking. As Dr. Kellerman asserts in her book, the leadership-training trend is somewhat nascent and based on the belief that anyone can become a leader if they take a few classes and apply a few principles. But as she points out, this does not seem to be necessarily true.[6] Max Van Manen, an educator and scholar, argues that the prescribed method for human science, in contrast to natural science, involves description, interpretations, and self-reflective or critical analysis. In other words, we explain nature, but we must understand human life.[7]

Scholarly work on leadership, in a relative sense, may be misdirected. We feel that the Newtonian approach to the study of leadership has been one of the reasons. Scientists have used the approach that if leadership can be broken down to its most basic unit or atomized, it can be rebuilt in a step-by-step method, thereby creating a prescription for building great leaders. Yet, because we are dealing with the dynamics of being human, leadership is a *subjective*, not an objective experience. Therefore, we subscribe more to an Einsteinian approach.[8] The Einsteinian view *infers* a reality based on the analysis of the relationship between the observed and the observer. It is easy to study, survey, quantify, and statistically analyze the formative and summative results of various leadership strategies to find what works. However, it does not tell us how it works, or how one leader is able to get results that another leader does not, while using the same strategies. Applying strategies or theories is only part of the leadership equation.

For many decades, if not centuries, scientific inquiry has been developed around mechanistic or Newtonian physics. Newtonian physics takes a somewhat deterministic or a linear view of cause and effect.[9] In other words, if one looks at a situation (we will call it situation #3), one should then be able to determine the evolution of situation #3 by looking backward to situation #2. Hence, situation #2 was a result of situation #1. The logical conclusion then, is that if one knows situations 1, 2, and 3, one can reasonably predict the results of situation 4, and so on.

In large measure, this is how the study of leadership has been undertaken and why there are so many differing theories. Researchers have identified successful leaders and tried to break down the individual components (traits, attributes, characteristics, etc.) to develop a strategy or recipe for creating other successful leaders. This approach is understandable because scientific inquiry is a strategy of trying to break things down to the smallest components in order to understand, quantify, and validate hypotheses and theories. Yet, is this the best method for studying the dynamics of human interaction?

Leadership and management require different strategies and competencies contingent upon the situations and their demands.[10] The question is, what competencies are needed for catastrophic or disaster crises? The term crisis denotes events that a) threaten high-priority values of people, organizations, or communities, b) exist in time pressures, and c) are unexpected. This existential event brings to bear good leadership or exacerbates the effects of poor leadership. It is for this reason a deeper look at leadership is needed.

First of all, what is the main function of management? If one really breaks it down to the most basic level, management is the process of controlling and compressing (to the best of one's ability) the events of one's environment in order to ensure efficiency and desired outcomes. Leadership, on the other hand, is expansive and nurturing. Leaders try not to be bound by firm policies, rules, and procedures because they may inhibit the ability to move toward a vision. Leadership's main role is to keep the vision (end) as the focus and foster a sense of community in seeking the vision. Management, while an important function in any operation, is very different from leadership. Management is the controlling activities of systems and things; leadership is the influence of relationships among people. In other words, we manage things, but we lead people. Since people are the real resource for getting things accomplished, we feel the need to focus more on leadership, particularly in the context of crisis events.

Before we introduce our LEAP model of leadership, which is a slightly different approach to developing leaders, we feel it relevant to give our definition of leadership. We posit that leadership is *a process and state where an individual influences a group, and the group agrees to the influence of the individual in order to reach a desired ideal or vision.* Notice the distinction from most definitions: *The group agrees to the influence of the individual in order to reach a desired ideal or vision.* We feel this an important concept to look at and one of the reasons why there is a lack of trust or support for leaders today. In a general sense, our leaders have taken a narcissistic approach to leadership, forgetting that the followers have to agree to their leadership. While the leader may enjoy a period of power with this view, their influence wanes very quickly and their leadership tenure is short-lived.

The LEAP Model

In light of the above information, we feel it relevant here to introduce a new leadership model. This model, designated by the acronym *L-E-A-P,* provides a guide for developing and assessing effective leadership and can be an effective tool in crisis leadership (or for that matter, any other discipline). We want to state up front, this is a model of leadership and not a leadership theory as such.

The LEAP model is composed of the four leader traits or characteristics we believe, through our observations and experience, to be most important. Therefore, a leader is:

- *L*egitimate: a leader has legitimate (authentic) power, knowledge, skill, and ability developed through his or her work, experience, education, and attention to detail;
- *E*thical: a good leader has strong character, strong values, and makes ethical decisions when confronted with choices and/or dilemmas;
- *A*ffective: a good leader has the ability to instill trust, confidence, emotion, passion, and create vision with others; and
- *P*ersistent: a good leader does not give up when times are tough or there is resistance to a righteous idea or plan, but rather shows determination in achieving goals and objectives.

Legitimate

Legitimacy is a social dynamic, which means that it is not a *thing* but a *relation* within a group of people. Importantly, each person's legitimacy requires at least one other to legitimize him or her in a social role. In fact, it is questionable as to whether or not an individual has any role at all, if there were no "others" for whom to play such roles. Roles are sometimes more concretely defined or else more situational in nature. Nonetheless, roles are legitimized by the others and ultimately are not an individual effort. In short, the leader is made legitimate by his or her followers and the followers acquiesce to the leader's power and influence. There is a relationship between leaders and followers that may not be a *thing*, but is nonetheless a reality for all of those within the relating group.

Leadership and management are two sides of the same coin and can be seen logically producing social outcomes. Søren Kierkegaard describes two antitheses of *possibility* and *necessity* that become synthesized in freedom. Possibility is the horizon of human potential and when used in this frame, we see that leadership is a cooperative intention of the group toward a goal or the horizon. On the other hand, management can be seen as the necessity or limitations on the group. Such limitations in organization can be economical, behavioral, communicative, etc. In other words, the rules, policies, and regulations in an organization are the confines of human interaction that make it predictable, sensible, and under control. Seen this way, it logically follows that a legitimate leader must have both expansive and creative qualities, as well as practical and mechanical sensibilities. The synergy of these qualities allows for a leader who is forward thinking, but can keep the group's motion on a meaningful and powerful path toward change.[11]

Ethical

Ethical leaders have been throughout history and currently are in great demand. Unfortunately, we have never been a morally perfect people, nor have we always honored the moral exemplar. However, it appears that more honor of the exemplar can be found in the past than now. This may be somewhat due to the fact that we have become a society that honors fame and popularity more than virtue. In fact, this is apparent with the high salaries of our professional athletes, actors, musicians, etc. Dr. David K. Hart asserts that the nation began to lose touch with its exemplars after World War II. Due to the rise of the big, complex organizations, individual accomplishment faded in with the cloudy background of collective behavior (group behavior). Therefore, by the 1980s it appeared as though the moral exemplars had become extinct. The problem is not that the moral exemplars are non-existent; it is that we as a society have lost the driving desire to find them and appreciate them when they are identified, which lessens their ability to influence.[12] It is within Dr. Hart's assertion we find an interesting paradox of today's society. Good leaders act righteously. However, this goes much beyond making choices based on a code of conduct or a code of ethics. It is actually a consistent demonstration of moral character; leaders act rightly in most, if not all they do.[13]

One of the things that made the United States such a successful enterprise was the concept of individual liberty. Each person was free to employ his or her talents and abilities to create the life he or she desired. While achieving this desired lifestyle may not have been easy, the liberty to pursue happiness was individual, not collective. The majority of society has held those who accomplish laudable or virtuous things in high esteem and while their success was created as a part of a community (and their accomplishments were, in part, aided by that community) it was through individual hard work and dedication that they were able to accomplish their goals. For this reason, many of our ancestors left everything they had to immigrate to the United States—to build a life for themselves and their families, where in large measure, they were in control of their destinies.

Yet, somewhere along the line, we have lost this concept. Individualism has taken a back seat to collectivism and we look at those who have accomplished laudable things differently than before. While we have been caught up in seeking to make everything fair and equal (which in and of itself is impossible), we have also recognized the need for a virtuous leader to provide the moral and principle-based guidance. We strive to find someone who will lead us to be better than we are. We want this to be universal in its application (collective), but it really is up to the individual. We are really looking for that courageous, virtuous human being who will exemplify the unique qualities we have heretofore held in such high esteem and help us to be better. However, the collective can only become better when the individual becomes better. This is what we believe everyone is secretly (or maybe even unconsciously) hoping for in a leader—someone who will lead us to be a better individuals.

So, what are the qualities of the ethical leader? Dr. Hart provides a great blueprint for the moral exemplar. He offers that the leader first must be of good moral character, must be a constant exemplar; it is not a sometimes thing. Second, the exemplar must not be compelled. He or she must be free to act as he or she intends. Third, the leader must be relatively faultless; he or she need not be perfect, but must continually strive for virtue. Finally, the actions of the leader must produce benefit or something good, even in failure; a leader's actions must never be pointless.[11] The attributes Dr. Hart has posited are timeless; they are not bounded by currency or locale. These characteristics have and will continue to be qualities of moral and ethical relevance. While modern moral issues may be different from past issues, these virtues and principles have, and can be anticipated to stand the test of time.

Affective

The ability to affect, or to be "affective," is very important in leadership. In many ways, it is the most recognizable attribute or characteristic of leadership. Yet, what does it really mean to be affective and how can one recognize a truly *affective* leader? In the opinion of many, a leader who affects is one who demonstrates charisma. Who does not want to be around a person who is charismatic? Unfortunately, charisma alone is not a good indicator of leadership and many cases exist where people have put their trust in a charismatic person, only to be led to harm or destruction.

We are born into this world as individuals; individuals with unique talents, abilities, personalities, weaknesses, and strengths. Most of what we do in life is to enhance ourselves individually. Yet, we as human beings are also very social. We seek to be accepted and interact with others. We are not only born with the ability to reason, we are also born with emotions and therefore are dynamic and diverse in our needs and desires. A leader who has the ability to affect has the power to create and transcend. We seek a leader with the ability to affect us and help us to feel accepted and secure, while at the same time helping us to effectively achieve and grow.

Effective vs. Affective

Unfortunately, due to the desire to make the most profit or benefit at the least amount of cost, there is dominant pressure to be as effective as can be. From a management perspective, effectiveness is very important and helps keep things working well. It is by no means a bad thing. Yet, effect without affect soon crumbles. Let's look at the difference between these two concepts.

Webster's Unabridged Dictionary defines *effect* as: *Anything brought about by a cause; a result.* With this in mind, we can deduce that to be effective is to bring about a desired result. Of course, this is a very important concept in individual lives as well as organizations. Everyone wants to bring about desired results, especially when one is implementing changes, growth, or seeking to maintain current situations. When something is working well and bringing desired results, we say that it is effective. On the other hand, *Webster's Unabridged Dictionary* defines *affect* as: *Influence, to stir the emotions of; to move.* Like with *effect,* we can also deduce that to be affective is to stir up one's emotion or provide a means for personal motivation to achieve desired results. While there appears to be subtle difference in the definitions, we believe the difference is much greater.

One of the prominent experts on leadership, Peter Drucker has provided some insight with his assertion *that management is doing a thing right, while leadership is doing the right thing.*[14] While this is somewhat nebulous or leaves a lot of room for interpretation, we feel it helps define what being affective means.

As we presented earlier, Van Manen asserted that human science, in contrast to natural science, involves description, interpretations, and self-reflective or critical analysis.[15] We define nature, but we must understand human life. Leadership involves human will and performance. We can quantify the results of leadership, but we will always have difficulty trying to define or quantify the process of leadership because it involves human interaction. Why? The answer is quite obvious: No two human beings are exactly alike. What works for one may not necessarily work for another. Also, what works in one situation may not work in a different situation. So how is it that we have been able to work together at all to achieve what we have? We believe the answer lies in the concept of *affectiveness.*

Vision

There is an old anecdote about a steamship crossing the ocean. Down in the boiler room, the firemen (those assigned to shovel coal into the boiler furnace) were being supervised by an effective manager. He had them working as hard as he could in order to make the best time possible. Meanwhile, the ship's navigator had been calculating their position and desired course. Upon realizing that they had made an error and were quite possibly lost, the navigator got hold of the supervisor of the firemen and asked him to stop shoveling the coal for a while because the navigator was not sure of where the ship was headed. The supervisor told the navigator to leave them alone because he didn't care where they were going, just that they were making good time.

This little story helps demonstrate the difference between *effectiveness* and *affectiveness.* The boiler room supervisor knew his role was to provide the most efficient and effective way of producing speed and he was doing his job well. Unfortunately, he had not brought to the boiler room crew the vision of where they were going. The only thing that mattered was to get the ship moving as fast as possible. While this story provides a little humor and one may think that this isn't realistic, the truth is it happens many more times than one would believe and is many times the reason individuals, families, and/or organizations fail. Generally speaking, we get so involved in the processes of life/business, we often fail to realize that we have no vision of our preferred future or desired destination.

So, what is a vision? The word comes from the Latin *videre,* "to see." What will the future look like? What do I see when I picture success? A vision provides a clear sense of purpose for an individual, group, or organization and provides a clear sense of future appearance (what the individual, group, or organization will look like or be) once the vision is realized. When a leader's vision is adopted or shared by the group, the group and individuals within the group tend to become empowered. Empowerment provides individuals with the capacity to believe, think, and act from a position of strength. The key element of empowerment is relationships, not accomplishments; relationships are more important. An empowered person will feel:

- important, legitimate, and significant;
- that his or her contribution and competence really matter;

- that his or her work is meaningful, exciting, and challenging; and
- that he or she has a sense of belonging to the group or organization.

A leader with a relevant vision can affect followers and create a pathway for individual and group success. In other words, an *affective* leader has his nose to the wind and his eyes on the horizon seeking to make things in the future more effective and efficient *through human interaction and relationships.*

Interpersonal Skills and Communication

Another attribute of the *affective* leader is his or her interpersonal skills and ability to communicate. In fact, we believe this is the major component or attribute that determines what many believe is charisma. Someone who can make others feel as though they are the only ones in the room has a great skill and we feel is the key to charismatic abilities. But, good communication is more than expressing one's vision and speaking with clarity. While these are definitely important and cannot be overlooked, communication is much, much more. As we have previously expressed, human relationships are more important than achievement or production for long-term success. Therefore, knowing and demonstrating how to interact with and communicate with individuals is extremely important. Even when one is speaking to a group, communication should be delivered as though speaking to an individual.

How can one become a better communicator—one who can *affect* others? The key is not in becoming more fluent and polished in one's speech (which is also important), but in becoming a better listener. An important thing to remember is that when one listens to others, one does not have to take on the other person's problems or try to solve them; if not, this is a mistake that can truly impede communication.

We would submit that listening involves four basic steps: (1) Listen—listen to what is being verbalized; listen to the emotions of the other person (anger, frustration, sadness, joy, etc.); (2) Understand—put yourself in their shoes to feel what they are feeling, understand what their needs are and why they are telling you these things; (3) Suspend all judgment—don't put your values or experiences in their situation; and (4) Do not offer advice—most people already know how to solve their issues and are communicating for validation.

If one is successful in incorporating these steps into communication, one will be much more successful as a leader, because followers will have confidence in the leader and will be empowered to accomplish more than they believe can. In many respects, this will help one to *affect* people in a positive way. One should be able to communicate a vision(s) in ways that will make others want to come along and be part of an exciting opportunity. This allows connection with others and empowers them to give their best efforts to become valuable members of a team—whether that team is part of the homeland security workforce or family.

Persistent

Much of what it takes to lead oneself and others is the tenacity to persist when the plan or process of achieving a goal becomes overwhelmingly difficult or seemingly impossible. An important aspect of persistence is the understanding that physical and/or political conditions can bring plans and actions to a halt. However, persistence is pursuing the desired outcomes in spite of perceived conditions and knowing only those things that are beyond our control or influence should cause us difficulty. Moreover, the more one knows about conditions in relation to his or her limits, the more influence an individual will have to adjust his or her path to success along the way. That means that effective leaders are more likely to actually build a greater capacity to withstand adversity. Life's challenges come in variations of mental, physical, spiritual, and combination tests of one's strength and endurance.

Staying the Course

Human beings have a paradoxical nature that has its nexus resting on anxiety. The anxiety we experience is grounded in the awareness of our own mortality. In other words, it is anxiety of dying that ultimately underlies our day-to-day worries, whether situational or general.[16] We understand that the old saying, "There are two things people have to do, die and pay taxes," is really only literally true for the first and not the second. Persistence in a challenging situation or condition means that we have to manage ourselves in the dance with our circumstances. All too often, we think of "fighting the challenge." The metaphor of a physical fight or war with our adversity can add psychological strain to the experience. Persistence and tenacity are not necessarily forcing through a barrier, but rather navigating through a field to reach a goal/solution, while also understanding that persistence is a process and only can be understood with respect to a desired outcome. This is why we often use metaphors of space and distance to talk about persistence. We can use persistence in terms of the metaphor of a war or of a journey.[17]

Focus

In the war metaphor, we end up dealing with issues like strategy, tactics, and destroying our obstacles. As a journey, life's challenging circumstance turn out to be obstacles, but we start thinking about different paths to take rather than how to blow them up. So, process is an important function toward outcome and is something we must attend to in terms of persisting in a task or project.

Susan Jackson and Mihaly Csikszentmihalyi assert that the relationship between a task's difficulty and the performer's competency is a ratio when, in adequate balance, provides an experience of absorption.[18] In other words, this is when the subconscious takes over so the mind and body function in total harmony. People will not experience this flow when the task is too easy relative to their abilities or when the task gets too difficult for their skill-sets. Therefore, we can see this generally in an individual's sense of self within the context of their situation. The principle applies whether it is an intellectual, physical, or spiritual challenge. When the task seems too big for the person, he or she will tend to find it overwhelming. When the task seems too small, he or she will take it for granted. When tasks are small, we can string them together to synthesize a larger task and find much more appreciation for its completion. In a sense, what we are proposing here is that in order to persist in a project or through a challenge without getting overwhelmed, bored, and/or losing focus, we have to mentally and emotionally "size" the experience to our capabilities. This means we have either to expand or constrict our worlds into a manageable relation in order to stay motivated and capable of meeting the task.[19]

Within the journey of persistence, one must remain focused on solving the various problems arising rather than fixating on the problems themselves. Moreover, the hyper-vigilance for problem detection and identification can be a drain on mental and emotional resources. Jason Selk calls the approach *relentless solution focus* that successful people must practice in order to see solutions to the problems rather than being *problem-focused,* which creates an overwhelming heap of problems. First of all, there are always challenges and problems that need to be navigated, but it is the spaces in-between that are pathways to the clearings and destination.[20]

When one is motivated by the outcome, without getting too focused on it, and is relentlessly solution-oriented rather than problem fixated, he or she can slalom through the barriers, obstacles, and uphill parts of the journey.[21] So, the pathway to success through persistence is to be goal oriented with enough flexibility and process mindedness to find the way through to the end. That means, one doesn't quit, but is stopped by only things beyond his or her control. One simply carries on with one choice, one action, and minding only those that are relevant to the process of the moment.

Summary

So where do we go from here? As we have pondered this for several years, one thing has become apparent to us: Leadership curriculum and scholarly activity need to undergo a paradigmatic shift. As we stated earlier, leadership is not prescriptive. We cannot send people to a class and present them with theories and strategies, which they can then attempt to apply in the real world. Leadership is a social, human science. First, we believe it to be counterproductive to atomize or break leadership down to its smallest components so that they can then be taught mechanically in leadership courses. Epistemology (what is known, theories, etc.) *is* important. However, the lived experience and human dynamic (ontology) is where true leadership flourishes.

Secondly, we believe a clearer distinction between leadership and management needs to be defined, studied, and put into curriculum. The model of director/teacher as leader has been used for years. Yet with newer technology and access to information, this model has apparently become less effective. While management skills are important and need to be continually taught, leadership is a differing concept and we may be obfuscating its importance by blending it with management. We think that the model of the teacher/student of the past has driven our acceptance of the leader/manager to be blended. If one is to look at business school curriculum at most universities, leadership is usually taught right alongside of management—often in the same course. It appears to be the assumption that when one rises to a position of management, that one is also a leader.

We believe that at the least, the curriculum at our universities should shift from teaching components of leadership theory to preparing leaders who have extensive work in personal development and have been given real-world experience in becoming leaders. We also believe that in addition to the current epistemological or theoretical aspects of leadership, the LEAP model is an effective guide for development of the human factors side of leadership curriculum.

To support our assertion, an interesting article was written by Erhard, Jensen, and Granger, titled: "Creating Leaders: An Ontological Model." They offered this in the abstract of their paper:

> The sole objective of our ontological approach to creating leaders is to leave students actually being leaders and exercising leadership effectively as their natural self-expression. By "natural self-expression" we mean a way of being and acting in any leadership situation that is a spontaneous and intuitive effective response to what one is dealing with.
>
> In creating leaders we employ the ontological discipline (from the Latin *ontologia,* "science of being," see Heidegger [1927]). The ontological model of leader and leadership opens up and reveals the actual nature of being when one is being a leader and opens up and reveals the source of one's actions when exercising leadership.[22]

One interesting assertion Erhard, Jensen, and Granger also make in this article points to what we have been saying throughout this book. They state:

> Yet, students being able to speak cogently about leader and leadership, and even with the ability to explain the style or characteristic, the principles and actions, and so forth that made or would have made this or that leader effective in this or that case, does not reliably leave students being leaders as their natural self-expression.[23]

In other words, learning the theories and components of leadership (epistemology), do not make one an effective leader. One must become or live the role of leader (ontology). For the most part current curriculum does not allow for that lived experience. We feel this is what is needed, especially with regard to our nation's security.

While it is obvious that formal education is not necessary to become a good leader or guarantee good leadership, it is in academia that most inroads or changes are researched and introduced. This is why we feel so strongly that a shift in paradigms be undertaken and new qualitative scholarly work done—we need to understand the human experience of leadership, not just the mechanisms or prescriptions of human interaction. As we come to understand more about that human experience, we will be better able to develop curriculum and programs to educate people better, who will then use their experience and development to become natural leaders—whether at home, in the community, in business, or in politics.

The LEAP model is a good basis to begin development of leadership curriculum and experience. We need to teach future leaders how to think critically, to find solutions to problems or roadblocks, and to become intransigent in seeking the righteous cause. Presenting future leaders with dilemmas and helping them to find innovative ways to succeed through reflection and experience will achieve this. With more quality leadership, we will raise the quality of living and provide better protection for the homeland, no matter the crisis. We can train managers, but we must *develop* leaders for the homeland security enterprise, and elsewhere!

Notes

1. Stephen R. Covey, *The Seven Habits of Highly Effective People* (New York: Simon & Schuster, 1989).
2. Robin Kielkowski, "Leadership during Crisis," *Journal of Leadership Studies* 7, no. 3 (2013): 62.
3. Barbara Kellerman, *The End of Leadership* (New York: HarperCollins Publishers, 2012), xv.
4. Martin C. Dillon, "Reversibility and Ethics," *Bulletin de la Société Américaine de Philosophie de Langue Française* 10, no. 2 (1998).
5. James B. Watson, "Psychology as the Behaviorist Views It," *Psychological Review* 20 (1913).
6. Kellerman, *End of Leadership.*
7. Max Van Manen, *Researching Lived Experience: Human Science for an Action Sensitive Pedagogy* (New York: State University of New York Press, 1990).
8. "Albert Einstein," *Brainy Quotes,* http://www.brainyquote.com/quotes/authors/a/albert_einstein_8.html
9. Barbara F. Okun, *Seeking Connections in Psychotherapy* (Hoboken, NJ: Jossey-Bass, 1990).
10. Montgomery Van Wart and Naim Kapucu, "Crisis Management Competencies: The Case of Emergency Managers in the USA," *Public Management Review* 13 (2011).
11. Søren Kierkegaard, *The Concept of Anxiety: A Simple Psychologically Orienting Deliberation on the Dogmatic Issue of Hereditary Sin,* trans. Reidar Thomte and Albert B. Anderson (Princeton, NJ: Princeton University Press, 1981).
12. David K. Hart, "The Moral Exemplar in an Organizational Society," in *Exemplary Public Administrators: Character and Leadership in Government,* ed. Terry L. Cooper and N. Dale Wright (San Francisco, CA: Jossey-Bass, 1992).
13. Ibid.
14. Peter F. Drucker, *The Essential Drucker* (New York: Harper Business, reissue edition, 2001).
15. Van Manen, *Researching Lived Experience.*
16. Ernest Becker, *The Denial of Death* (New York: Free Press, 1997).
17. George Lakoff and Mark Johnson, *Metaphors We Live By* (Chicago: University of Chicago Press, 2003).

18. Susan Jackson and Mihaly Csikszentmihalyi, *Flow in Sports: The Keys to Optimal Experiences and Performance* (Champaign, IL: Human Kinetics, 1999).
19. Kirk J. Schneider, *The Paradoxical Self: Toward an Understanding of Our Contradictory Nature* (New York: Humanity Books, 1999).
20. Jason Selk, *10-Minute Toughness: The Mental Training Program for Winning before the Game Begins* (New York: McGraw-Hill, 2008); Jason Selk, *Executive Toughness: The Mental-Training Program to Increase Your Leadership Performance* (New York: McGraw-Hill, 2011).
21. Kay Porter, *The Mental Athlete: Inner Training for Peak Performance in All Sports* (Champaign, IL: Human Kinetics, 2003).
22. Werner Erhard, Michael C. Jensen, and Kari L. Granger, "Creating Leaders: An Ontological/Phenomenological Model," in *The Handbook for Teaching Leadership*, ed. Scott Snook, Nitin Nohria, and Rakesh Khurana (Thousand Oaks CA: Sage Publications, 2012).
23. Ibid.

10

Ethical, Legal, and Social Issues (ELSI) in Homeland and Civil Security Research and the European Union Approach

KATERINA HADJIMATHEOU, TOM SORELL, & JOHN GUELKE

Introduction

In this chapter, we introduce the main ethical, legal, and social issues that arise in connection with homeland and civil security research anywhere, and the methods used in the European Union to address them. In the *first section*, we discuss some of the challenges for governments that arise from the need to balance liberty and security in the face of threats from crime and from emergencies. In the *second section,* we consider the ethics of setting the security research agenda, research ethics for homeland security, and specific ethical, legal, and social issues arising with the use and export of security technologies. In the *third section,* we introduce some of the technical and regulatory means used to address the ethical, legal, and social concerns raised in the second section, including privacy by design, national oversight mechanisms, training, codes of conduct, and export and licensing regulations. Finally, in the *fourth section* we provide an overview of the European Union's approach to incorporating ethical, legal, and social concerns into the security research process.

Balancing Liberty and Security

States have duties to protect the security of citizens. They also have duties to guarantee citizens' rights and freedoms. Basic rights are stated in both international and domestic constitutional law. Basic rights are usually freedoms that enable people to lead lives without undue interference or restriction by governments. International law—in the form of human rights treaties—and the domestic constitutional law of liberal states leaves people free to lead private lives, enter professions of their choosing so long as they are appropriately qualified, and to participate peacefully in public life, for example by voting, joining a political party, or becoming a member of a union. Both constitutional and international human rights law constrain governments to refrain from legislating in ways that restrict freedom or that treat people unequally.

Ordinary criminal legislation in a liberal jurisdiction threatens loss of liberty to people who are legally proven to have committed certain offenses, usually violent offenses or offenses against property. Domestic constitutional law and international human rights law often limit the severity of punishment even when the commission of crime has been proven, and they constrain the procedures that may be followed by the state in collecting evidence against offenders, in arresting and holding suspects, and in prosecuting them for crimes through trials. For example, capital punishment has been abolished in Europe, and European law forbids the extradition of people to capital punishment jurisdictions, if there is a chance that these people would be tried in those jurisdictions for capital offenses. This ban on extradition includes extradition to jurisdictions in which due process is observed, such as those of states of the U.S. that inflict the death penalty.

Protecting people's security sometimes involves limiting the freedoms of a whole population. So long as the operation of these limitations is kept as short as possible and imposed in order to protect important rights, such as the right to life, even human rights law permits them. Human rights law recognizes the existence of emergencies that "threaten the life of the nation." Both the *European Convention for the Protection of Human Rights and Fundamental Freedoms* (1950) and the *International Covenant on Civil and Political Rights* (1966) recognize this kind of emergency. Anticipating that declarations of emergency might be used opportunistically by governments as justification for the unnecessary limitation of rights, human rights law discourages the declaration of an emergency by governments, and requires the period of emergency to be as short as possible. Even in emergencies, certain human rights may not be limited, according to human rights law. These include the right not to be tortured and the right not to be discriminated against.

In human rights law, the central case of an emergency is an internal or external war, or, if it is big enough, an imminent terrorist attack. Emergencies in the sense of natural disasters are less central. State responses to security threats posed by emergencies often involve interfering with people for their own good. For example, sometimes governments operate forced evacuations to protect people from flooding or contamination. State responses to political violence and terrorism often involve interfering with some people—those suspected of criminal intent—for the protection of others. In both kinds of cases, a balance between security and liberty must be struck. Some of the balancing is reflected in limitations clauses of human rights instruments. For example, free travel may be abolished in a period of emergency, and the right to free speech is supposed to be limited by the need not to damage unduly people's reputations. Another way in which the need for balance can be reflected is in time-limiting emergency powers. In many jurisdictions, emergency powers last only for a period of weeks before a legislature has to renew them. Analogously, the period in which anyone can be held without charge by a government, even when suspected of a terrorist crime, is usually restricted legally in certain jurisdictions. In addition to time-limiting certain measures, constitutional and human rights law sometimes demand that governments apply to courts for permission to implement measures. For example, governments and police forces in liberal states usually have to apply with good reasons to a court in order to install a wire-tap, conducting a search of a house, or mounting a surveillance operation that consists of following suspects as they go about their business.

Ethical, Legal, and Social Issues Arising from the Research, Development, Use, and Sale of Security Technologies

Constitutional and international human rights law apply primarily to security measures taken by states. But they also have a bearing on security technologies produced within a state, in particular security technologies developed with state funding by state-owned technology companies. Some state-owned or state-funded companies produce weapons, for example. Some of these are for military use; others might be sold to police forces. International agreements limit the kinds of weapons that can be produced. Chemical weapons are universally banned, for example. And it is internationally agreed that the number of producers of nuclear weapons should be limited. But there are well-recognized ethical issues raised by the sale of arms and military equipment to belligerent, undemocratic states, especially when these arms are being used or could probably be used against local populations. The most belligerent states are sometimes subjected to trade sanctions that cover the export to them of military equipment or materials for weapons. Different but still difficult issues arise for non-lethal security technologies—such as surveillance equipment, detection technology, and equipment that facilitates data collection from a local population.

Ethical, legal, and social issues arise at all stages of the security research process: from the development of the research agenda as a whole, the design of individual projects, and the execution of the research, to its eventual application in the real world. The issues include questions about the imposition of export restrictions and perhaps also the regulation of secondary markets in which technology is re-sold. In this section, we provide an overview of all of these issues, distinguishing between those that arise at different phases.

The Security Industry and the Governance of Security Research

Recent years have seen a rise in public and political expectations of the security industry to provide emergency services, police, and intelligence officials with technologies that can help them prevent and respond to security threats. The range of technologies that have been and are now being developed is vast, and includes smart closed-circuit television (CCTV) cameras; biometrics; UAVs for surveillance; border security and crowd control technologies; and communications monitoring, to name just a few. This reflects the expansion of the market for such technologies, the pace of which is remarkable: In 2011, the *Wall Street Journal* reported that "the annual value of the retail market for surveillance has increased from 'nearly zero' in 2001 to around $5 billion." In the EU, the budget for non-military security research in the *7th Framework Programme* was €1.4 billion; for the successor *Horizon 2020* program, it is €1.695 billion.

In the EU, concerns have arisen in relation to the way the European Commission security research agenda is developed, and in particular the interests that are represented by those setting the agenda. Most of this research is commissioned. That is, topics are decided by the European Commission in conjunction with advisors from the Member States, industry, and academia. A succession of reports scrutinizing the EU security research program criticized what they saw as the excessive and largely hidden influence on the security research agenda of the very corporations that most stand to gain from security research funding:[1]

> Within the 660 "stakeholders" participating in the 11 working groups established for the purpose of drafting the ESRIF [European Security Research and Innovation Forum] report, Statewatch researchers find that 66% (433) come from the defence and security industry (ASD, EADS, Finmeccanica and Thales [all large security companies] account for 102 participants), 30% (200 participants) from national and European security services and agencies, and only 1.4% (9 participants) from "civil society," out of whom no representative from a civil liberties or privacy organization.[2]

In 2013, a follow-up report on the same topic was commissioned. It found that the same interests continue to be overrepresented and that research topics tend to focus on enhancing the effectiveness of technology while "sidestepping issues linked with their legitimacy." The report concluded that the "absence of ethical reflexion on the uses of technologies of digital surveillance, in particular the impact that these technologies can have on the rule of law is particularly striking."[3]

Research Ethics for Security Research

Research ethics is a set of norms applying to scientific research in general. It aims to regulate the way in which research involving human beings is carried out. The principles that apply to security research are the same as those that apply in all other fields and include: consent of participants; respect for confidentiality and data protection; and avoidance of conflicts of interest, among others. Some security research involving human participants easily lends itself to traditional methods for meeting the requirements. Information sheets can be distributed and consent forms signed at airports, for example. But what about large-scale public gatherings where emergency-response equipment is being tried out, or field-testing for CBRN-E technology, where the "volunteers" might be serving members of the emergency services who cannot easily say no? How is technology that is designed to be deployed secretly to be tested in conditions in which people are aware it is being used on them and agree to it? These are some of the questions that research ethics boards face when considering applications for funding.

Ethical, Human Rights, and Societal Issues around the Implementation of Security Research

The use in practice of security systems and technologies raises a variety of legal, social, and ethical issues. For example, human rights scholars and activists have pointed out that the use of security technologies poses threats to individual rights, including privacy, data protection, and non-discrimination as well as to freedom of expression, association, movement, and religion, among others. Criminologists and surveillance studies scholars have highlighted the social issues raised, such as the potential impact of such technologies on disadvantaged groups, on community-police relations, on trust among citizens, and on public perceptions of security. As ethicists, we are concerned with all of these issues, because they all have a bearing on whether and when the use of such technologies is legitimate in practice. In this section, we present the main legal, social, and ethical issues raised by security technologies, grouping them according to the controversial aspects of their uses and features.

Dual Use of Technologies

Security technologies are tools and, like all tools, they can be put to bad or morally questionable purposes. Some of these purposes are ethically acceptable; others are deeply problematic. Airborne surveillance systems and UAVs can be used to detect and map oil slicks. Such detections can help the authorities to contain environmental damage and reduce the costs of clean-up. But they have also been used by border forces to identify vessels full of unwelcome migrants, which they then "push back" into international waters or into the territorial waters of other countries. This kind of "pushing back" denies migrants who were intending to seek asylum abroad their legally and internationally protected human right to do so. It has also delivered migrants back into the hands of people or regimes who intend to violate their rights. Marine border surveillance technologies can be used both in ways that protect basic human rights and in ways that violate them.

Crowd control technologies for location tracking and communications monitoring have also been used in rights-protecting and rights-violating ways. They can be used to help police respond to riots to protect people, shops, and property from damage. But, as the example of the Arab Spring illustrates, they have also been used by authoritarian regimes to identify and monitor people who attend political protests, in order to arrest, interrogate, and even torture or kill them.[4] At the very least, this latter use of such technologies violates people's rights to express their political views and to associate with other people in order to do so.

Mission Creep

Mission creep is the realization of dual use over time. It occurs when a tool or technique is used not only for its original purpose (or "mission"), but for other, sometimes unrelated purposes. In a democracy, the use of security technologies is itself legitimate only when it has been approved by a legitimate authority (such as the legislature) as necessary and proportionate to the pursuit of a legitimate goal. The fact that a specific application of a particular technology has been authorized in this way does not mean that people can no longer rightly object to it or that it is justified, all things considered. But, it does mean that one basic and important condition of legitimacy has been met. When a technology is authorized for one use, but then applied without authorization to another, this basic condition has not been respected. The legitimacy of the use of technology for this new purpose is therefore questionable. For example, in the UK, ANPR (automated number-plate recognition) cameras were first installed in order to deal with car crime. Today, they are used for a wide range of policing purposes, including counterterrorism and crowd control. These new uses have been introduced incrementally, without statutory regulation.[5] As a result, the intrusions into privacy and restrictions on freedom of movement imposed by ANPR are considered by many people to be illegitimate.

Human in the Loop

Ethical issues arise with both the inclusion and exclusion of humans in the process of applying security technologies. Let us consider first those that arise in connection with the fact that technologies are operated and implemented by humans. Humans are psychologically and cognitively imperfect in a range of ways that affect the effectiveness of technologies and the fairness of their application. For example, studies have shown that humans can pay attention to events on a CCTV monitor for no more than about 20 minutes, after which their ability to spot real-time security threats effectively dives dramatically.[6] As is discussed below in more detail, greater automation has been proposed as one means of reducing the presence of the human in the loop and thereby mitigating these risks. Automation is unlikely to help correct plain old human errors, however. Humans make mistakes, which can harm people if they lead to the misidentification of a suspect or the mistaken categorization of a person as suspicious or dangerous. For example, errors in the typing, spelling, or translation of names has led to some people being mistakenly included on watch lists and as a result being prevented from travelling.[7] They also include breaches of data security resulting from sloppy or lax practices.

Automation brings its own set of ethical and legal risks. Accountability for actions is a particular concern. Accountability requires that decisions that have an impact on individuals' rights and freedoms should be traceable to an individual or a body of people authorized to make that decision and who can be held responsible. It also requires that the effectiveness and proportionality of an action should be verifiable after the fact. A data mining technology that uses extremely complicated algorithms to analyze huge quantities of information from a variety of sources to determine who should be singled out for surveillance may raise issues of accountability if no single individual is able to verify the reliability of that determination. In addition, sometimes technology may be insufficiently sensitive to morally significant distinctions to deploy force proportionately. In such cases, human judgment may be required before action can be taken legitimately.[8] For example, a new ship-mounted technology called WatchStander can detect and target "countermeasures"—very bright lights designed to confuse—at small vessels, often used by pirates and smugglers in order to evade ship radars.[9] A problem arises because the technology detects all small vessels but cannot distinguish between those used by pirates and those used by irregular migrants and refugees, who also use small fishing boats to cross maritime borders. Targeting frightening and disorienting lights at these boats poses a threat to the security of those on board. It may lead the pilot or the passengers to panic and cause the boat to capsize. It may even prompt the smuggler responsible for the vessel to intentionally cause it to capsize in order to make the crew of the larger ship mount a rescue operation, averting attention from their own criminal actions. Unless and until the technology can distinguish between pirates and migrants, human intervention should be required between detection and the deployment of countermeasures.

Risks of Profiling: False Positives and Discrimination

Profiling involves the reliance by authorities on generalizations about the danger posed by people who display certain traits, characteristics, or behaviors. Because profiling involves generalizing, it is inevitable that some of those who fall within or match a profile will not in fact be suspicious or dangerous. Identifications of these people by a profile are called "false positives." The number or proportion of false positives created by a profile reflects on its acceptability as a tool, because despite their innocuousness, the people identified have to bear whatever costs or interference the profile imposes. The broader the profile and the weaker the evidence on which it is based, the greater the likely proportion of false positives and the harder it is to justify.

Ethnic profiling involves the use of ethnic traits (typically by police, but also potentially by emergency services) as an indicator of the need for further investigation. Ethnic profiling is controversial because it has been used in ways that associate certain ethnic groups with certain kinds of criminality. For example, following the 9/11 terrorist attacks of 2001, the German government adopted a program of data mining of government and social services records to identify male students of Muslim background.[10] The German Federal Court later outlawed the policy, citing the stigmatization of Muslims as one of the reasons against it. Ethnic profiling can encourage people to perceive certain groups as ill-intentioned and dangerous and can lead to discrimination against them. It can also lead to a "ratchet-effect" whereby the more certain groups are prioritized, the greater the numbers from those groups represented in the criminal population, the stronger the statistical rationale for targeting those groups.[11] Similar problems have been identified with the use of profiling by police to identify which areas in a city are more likely than others to be to be sites of criminality.[12]

Discrimination-related problems also arise when individual profilers—such as police or border guards—are permitted to use their own judgment of what is suspicious to define the profile they apply. The quality of human judgment is affected by a range of influences, including prejudice, susceptibility to pressure from others, and confirmation bias (a tendency to favor or give excessive importance to evidence that confirms one's pre-existing suspicions or beliefs). Studies have shown that these factors have influenced detrimentally the way in which scientists interpret forensic evidence used in criminal trials,[13] and the way in which police conduct investigations.[14]

Privacy

Privacy is very often at risk from security measures, especially those aimed at reducing crime and terrorism. Intrusions into privacy are an inevitable cost of counterterrorism measures, because terrorists inevitably work in secret, concealing their activity within the sphere of privacy usually granted to those who are harmlessly pursuing family life or friendship. Police efforts to acquire information involve gaining access to environments where conspirators speak relatively openly, and these include private environments.

Privacy is valuable for many reasons, two of which are particularly worth highlighting here. The first relates to the fact that without privacy people cannot form and maintain relationships that are vital to their well-being.[15] For example, the quality of friendships, romance, parenting, and some kinds of professional relationships depends on people's ability to exclude others from their sphere. Monitoring of communications that intrude on relationships is considered particularly intrusive for this reason. The second reason privacy is valuable relates to the value of democracy: The quality of democracies depend on people's freedom to question accepted views, flirt with minority or controversial ideas, and to debate with and persuade each other of the rightness of different views. People will not feel free to engage in such activity if they know they are being watched and potentially judged while they do. For this reason, monitoring of political activity is also considered particularly intrusive of privacy.

Data Protection

The right to have one's data processed (meaning stored, communicated, and analyzed) securely and in line with one's other rights, especially the right to privacy, is recognized in EU jurisdictions but not in the USA. Article 8 of the *Charter of Fundamental Rights of the European Union* states that personal data must be processed fairly for specified purposes and on the basis of the consent of the person concerned or on some other legitimate basis laid down by law; that everyone has the right of access to data that have been collected concerning him or her, and the right to have it rectified; and that compliance with these rules shall be subject to control by an independent authority. Data protection rights overlap with, but are not subsumed by the right to privacy. Privacy intrusions often result if unauthorized people have access to data. But not all privacy intrusions involve the collection or processing of data. And some processing of data will violate the right to data protection even when the data is openly available and no one's privacy has been invaded.

The Chilling Effect and Trust

The *chilling effect* occurs when people become reluctant to take part in certain activities because they fear participation will bring them under official suspicion of criminal activity. Chill is problematic when it leads people to opt out of activities that are legitimate and desirable in a free society, such as free association, free speech, and political organization. An exhaustive account of how comprehensive monitoring of telecommunications by an authoritarian regime has chilled political activity is provided in a recent report on Ethiopia by the campaign group Human Rights Watch.[16]

It has been claimed that, by encouraging people to perceive others as the source of security threats, the use of security technologies also discourages people from trusting one another.[17] *Trust* helps people feel able to rely on others in ways that enable them to pursue their own interests more effectively. Economic and political theorists have shown convincingly how important interpersonal trust is for a country's political stability and economic success. Security technologies used to single out certain groups of people as potentially dangerous may discourage others to trust and interact with them. This may lead to the social exclusion of those groups, to their general detriment. Security may therefore have to be balanced not only against liberty but also against risks to trust—interpersonal trust between citizens in general or trust of some citizens in the goodwill of others.

Security Paradox

If surveillance practices draw attention to the presence of crime so that people begin to overestimate the threat it poses, then this may increase feelings of insecurity (in the form of vulnerability to security threats). This phenomenon has been termed the *"security paradox."*[18] Some recent country-specific research suggests this paradox does not occur in relation to, for example, the use of open-street CCTV, but it is difficult to know how far we can extend these findings to other countries and the use of other surveillance practices.[19] Another kind of security paradox may arise from security measures that are used in ways that are perceived as unfair by those they affect. These can have the unintended effect of decreasing sympathy for the state and increasing sympathy for violent causes, thus aggravating counterproductively the security threat. This occurred in Northern Ireland, where a range of heavy-handed security measures undermined Catholic confidence in the British state authorities.[20] Finally, it has been proposed that a related paradox of security measures, referred to as the *moral hazard,* might arise. Moral hazard may occur because the more secure people feel when engaging in certain activities, the more recklessly they behave, thereby increasing the risk to themselves of harm. Concerns about moral hazard have led some to worry that additional security measures can lead to diminishing real security.[21]

Preempting, Managing, and Mitigating the Ethical, Legal, and Social Risks of Security Research

This section provides an overview of some of the measures that are taken to address the legal, social, and ethical risks identified above. It begins by describing some measures taken by those who research and develop security technology. It then considers what the users and regulators of such technologies can do to protect ethics and rights.

Privacy by Design, Privacy-Enhancing Technologies, and Privacy Impact Assessments

Sometimes, security technologies are developed that are so problematic ethically that their use is unlikely ever to be acceptable in practice. For example, airport body scanners such as Smith's millimeter wave model reveal to operators the naked bodies of the individuals who pass through them. Bodily privacy is too important to most people to be worth compromising in this way. The release of the scanner onto the market met with serious public and political opposition. Its use has now been banned in the EU. Fortunately, there are things security researchers and developers can do to avoid producing a tool that is so ethically, legally, or socially unacceptable that it is unmarketable.

Privacy by design is the practice of building privacy protecting features into security practices and technologies to preempt violations of privacy. For example, a privacy-by-design technique was used to produce socially, legally, and ethically acceptable body scanners: Instead of producing an image of the naked body, these scanners show operators a simple line-drawing of a generic human body, and represent the location of any suspect object by means of a flashing light. This simple solution preserves the detection function of the technology while minimizing the intrusion into bodily privacy. Privacy-by-design principles that now inform technology development include data minimization, which requires that technology is developed so as to minimize both the amount of data stored, shared, or analyzed and the length of time for which it is kept. This mitigates against mission creep as well as privacy intrusions. *Privacy by default* refers also to technological features like anonymization and encryption, which make it harder to identify individuals via data and which mitigate the privacy impact of security breaches. Automation can also protect privacy by reducing the numbers of humans who actually look at information. *Accountability by design* refers to efforts to prevent misuse of technologies and ensure public accountability by designing automatic log history into systems. Disabling features can also be built into technologies to enable manufacturers or authorized users to enable them to halt the function of a system that is being misused.

Privacy impact assessments are management tools that organizations and companies can use to help them protect people's privacy and rights to data protection when undertaking projects that involve the processing of personal data. They invite organizations to think about the way they process data and to put in place measures to ensure safe, transparent, accountable, and lawful processing of data. Privacy impact assessments advocate a privacy-by-design approach to developing ways of using people's information. They are, however, time-consuming and bureaucratically onerous, which may discourage their application, especially by smaller organizations.

Codes of Conduct and Training of Technology End-Users

Error, misuse, security breach, and problems of prejudice and discrimination can all be addressed through training of technology users, as well as the adoption of codes of conduct guiding the use of technologies. Officials can be trained to spot insider threats, to whistle-blow, and to be aware of their own tunnel vision, confirmation bias, or prejudices. Codes of conduct that encourage and enshrine the rights of whistleblowers can help those applying security measures to feel supported in reporting poor or unethical use of technologies. Codes of conduct that demand respect for the rights of those affected by security measures provide useful opportunities for end-users to resist any pressure by governments to use security technologies in ethically questionable or human-rights infringing ways. For example, as noted above, border guards who are under political pressure to reduce illegal migration may prioritize this over the protection of the human rights of migrants. If they operate under a code of conduct that enshrines the protection of human rights as a primary duty, they can point to and emphasize this in order to resist such pressure.

National Security and Information Oversight Mechanisms

The legal and social impact of measures of homeland and civil security are in many jurisdictions subject to oversight mechanisms, including:

- parliamentary committees that examine and report on implementation of security powers and techniques (an influential example from the UK);
- independent appointees such as rapporteurs or "independent experts" that monitor and report on specific laws or powers; or
- information commissioners or data protection authorities that monitor the effects on data protection and privacy rights of all kinds of practices, including security practices. These authorities often also hear complaints and provide redress for individuals who feel their rights have been infringed.[22]

Licensing and Export Controls

Licensing and export controls are two means by which governments try to control sales of security technologies produced in their jurisdiction. The aim of licensing is to ensure that technologies are only used by individuals, companies, or agencies that can demonstrate that they need to use them to fulfill a pressing and legitimate purpose. Export controls are an instrument enabling governments to monitor and control the kinds of technologies sold abroad by domestic companies to ensure they comply with foreign policy objectives as well as human rights obligations. Awareness of the shortcoming and weaknesses of export controls, in particular the failure to keep pace with changes in technology,[23] led to the establishment of the Wassenar Arrangement.[24] This is an international effort by 41 nations, including the major exporters of security technologies, to establish common binding rules on exports of security technologies.

Building Ethical, Legal, and Social Considerations into European Security Research: The European Commission Approach

In this final section, we address the role of those who fund security research in addressing ELSI. Specifically, we introduce and reflect on the European Commission approach to building ethical, legal, and social considerations into the security research process.

For the period 2014–2020, the European Commission has allocated €1694 million of public funds to the "Secure Societies" research stream.[25] Public and political concerns with the ethical, legal, and social impact of security research in the EU has led to sustained efforts by the Commission to address these issues at all stages of the research process.

The European Commission's Directorate General for Home Affairs (DG HOME) is now the government body that administers EU-funded security research that was previously within the remit of the Directorate General for Enterprise and Industry (DG ENTR). Their official call (invitation, specifying research topics and objectives) for security research funding now explicitly requires those applying take account of ethical, legal, and societal issues in their projects. Sometimes this requirement is met by establishing an independent board of ethics and law experts who read, comment on, and hopefully influence, the research, so that it is in line with liberal democratic laws and values. Sometimes it means including within the project a stream of research (known officially as a "Work Package") undertaken by academics with relevant expertise in law and ethics. This Work Package usually identifies all relevant legal, social, and ethical issues arising from the products or techniques being developed and makes recommendations. In addition to this, all applications for security research are vetted by a European Commission-appointed ethics board and all must obtain approval from authorized Ethics Research Committees prior to starting their work.

Along with projects developing security technologies, the EU funds, under the security topic, projects analyzing public attitudes to security and projects that analyze the ethical, legal, and social aspects of security research and security policy in the EU. These projects often involve collaboration between academics from fields such as law, philosophy, and social science and security practitioners such as police forces (relevant EU security research projects include SURVEILLE,[26] DETECTER,[27] IRISS,[28] and RESPECT[29]). It is intended that the work of these projects will affect the policy-making process, leading to better security policy, regulation, and funding decisions in the future. As part of the SURVEILLE project, the European Commission also funds an ethics and fundamental rights advisory service, which technology developers can consult for confidential advice on how to ensure their research is compatible with ethical and human rights standards.[30]

In 2010, DG ENTR established a Societal Impact Expert Working Group (SIEWG) composed of experts from the security industry, academia, NGO, and policy sectors, to act as an "early warning system for undesirable impacts" of systems and technologies proposed for research. In 2012, the SIEWG reviewed measures already in place at the EU level to take account of and manage the societal impact of security research and made recommendations addressing each stage of the research process. One of their overarching recommendations was that there should be a greater involvement of civil society groups both in the design of the research program and at the funding and review stages. Another was that societal impact considerations are broad and require initiatives that go well beyond privacy-enhancing technologies and data protection. Although this has not been considered by the SIEWG, there may be a case for large-scale interactive events involving project leaders and interested politicians and members of the public.

The SIEWG recommended that greater focus be given to societal issues at the very early stages of the setting of the security research agenda, especially by the Security Advisory Group, a group of 21 independent experts drawn from security end-user organizations, academia, and industry with three representatives from the European Commission, which shapes the EU security research call for proposals. In particular, they recommend that the Security Advisory Group include greater representation from civil society groups. They also consider the societal relevance of the proposed research; the potential impact on citizens' rights; and its necessity and proportionality. According to the SIEWG, research is socially relevant when it is "relevant to the needs of society, addresses their fears and provides understandable solutions."[31] The SIEWG further recommended strongly that EU bureaucrats working in the administration of security research undergo education and awareness training in relation to the societal impact of security technologies and processes. The aim of such education is to help those responsible for administering and overseeing security research become more sensitized to ethics and human rights issues and more able to identify and take action in response to them.

Notes

1. Ben Hayes, "Arming Big Brother. The EU's Security Research Programme," *Transnational Institute Briefing Series,* no. 1/2006, http://www.statewatch.org/news/2006/apr/bigbrother.pdf; Ben Hayes, *NeoConOpticon: The EU Security-Industrial Complex* (N.p.: Transnational Institute in association with Statewatch, 2009), http://www.statewatch.org/analyses/neoconopticon-report.pdf; European Commission, Directorate General for Internal Policies, Policy Department; Citizens' Rights and Constitutional Affairs, *Review of Security Measures in the 7th Research Framework Programme: A Study*. 2010, http://www.statewatch.org/news/2010/nov/ep-review-security-research-programme.pdf
2. European Commission, *Review of Security Measures in the 7th Research Framework Programme,* 15.
3. European Commission Directorate General for Internal Policies, Policy Department; Citizens' Rights and Constitutional Affairs, *Review of Security Measures in the 7th Research Framework Programme: A Study,* 2014, 7, http://www.europarl.europa.eu/RegData/etudes/etudes/join/2014/509979/IPOL-LIBE_ET(2014)509979_EN.pdf
4. "The Surveillance Industry Index," *Privacy International,* November 2013, last published May 23, 2014, https://www.privacyinternational.org/?q=node/403
5. Alan Travis, "Big Brother Traffic Cameras Must Be Regulated, Says Home Secretary," *The Guardian,* July 4, 2010, http://www.theguardian.com/uk/2010/jul/04/anpr-surveillance-numberplate-recognition
6. Kevin Macnish, "Unblinking Eyes: The Ethics of Automating Surveillance," *Ethics and Information Technology* 14, no. 2 (2012).
7. Mike McIntire, "Ensnared by Error on Growing US Watch List," *New York Times,* April 6, 2010, http://www.nytimes.com/2010/04/07/us/07watch.html?pagewanted=all&_r=0
8. Home Office (UK), *Surveillance Camera Code of Practice* (London: The Stationary Office, 2013), https://www.gov.uk/government/uploads/system/uploads/attachment data/file/204775/SurveillanceCameraCodeofPracticeWEB.pdf
9. Hal Hodson, "Pirates Incoming! Ship Radar Keeps Watch and Hits Back," *New Scientist,* May 30, 2014, http://www.newscientist.com/article/mg22229714.700-pirates-incoming-ship-radar-keeps-watch-and-hits-back.html#.U42qLUJidUQ
10. Daniel Moeckli, "Discriminatory Profiles: Law Enforcement after 9/11 and 7/7," *European Human Rights Law Review* 5 (2005).

11. Bernard E. Harcourt, *Profiling, Punishing, and Policing in an Actuarial Age* (Chicago, IL: Chicago University Press, 2006).
12. Elizabeth E. Joh, "Policing by Numbers: Big Data and the Fourth Amendment," *Washington Law Review* 89, no. 35 (2014).
13. John B. Gould, Julia Carrano, Richard Leo, and Joseph Young, "Predicting Erroneous Convictions: A Social Science Approach to Miscarriages of Justice," research report submitted to the U.S. Department of Justice, February 2013, https://www.ncjrs.gov/pdffiles1/nij/grants/241389.pdf
14. Keith A. Findlay and Michael S. Scott, "The Multiple Dimensions of Tunnel Vision in Criminal Cases," *Wisconsin Law Review* 2 (2006).
15. James Rachels, "Why Privacy Is Important," *Philosophy and Public Affairs* 4 (1975).
16. Human Rights Watch, *They Know Everything We Do* (N.p.: Human Rights Watch, 2014), http://www.hrw.org/reports/2014/03/25/they-know-everything-we-do
17. Jonathan Kimmelmann, "The Promise and Peril of DNA Databanking," *Nature Biotechnology* 18 (2000); David Lyon, *The Electronic Eye: The Rise of the Surveillance Society* (Minneapolis: University of Minnesota Press, 1994).
18. Societal Impact Expert Working Group, "Report of the Societal Impact Expert Working Group," EC DG ENTR report, 2012, http://cies.ie/wp-content/uploads/2014/05/Report-of-the-Societal-Impact-Expert-Working-Group.pdf
19. Belgian Ministry of the Interior, "Video Surveillance in Public Spaces—Is It Really Efficient?" Study of the Katholieke Hogeschool Zuid-West-Vlaanderen, Expertisecentrum Maatschappelijke Veiligheid for the Belgian Ministry of the Interior, Flemish version, June 2013.
20. Michael P. O'Connor and Celia M. Rumann, "Into the Fire: How to Avoid Getting Burned by the Same Mistakes Made Fighting Terrorism in Northern Ireland," *Cardozo Law Review* 24 (2003).
21. Societal Impact Expert Working Group, "Report of the Societal Impact Expert Working Group," 4.
22. For a report on provisions for redress in the EU see Fundamental Rights Agency of the European Union, *Access to Data Protection Remedies in EU Member States* (Luxembourg: Publications Office of the European Union, 2013), http://fra.europa.eu/sites/default/files/fra-2014-access-data-protection-remedies_en_0.pdf
23. "The Surveillance Industry Index," *Privacy International.*
24. "Wassenaar Arrangement on Export Controls for Conventional Arms and Dual-Use Goods and Technologies," Wassenaar Arrangement on Export Controls for Conventional Arms and Dual-Use Goods and Technologies, Secretariat, last published February 12, 2015, http://www.wassenaar.org
25. See "HORIZON 2020. The EU Framework Programme for Research and Innovation: Secure Societies—Protecting Freedom and Security of Europe and its Citizens," European Commission, last accessed March 6, 2015, http://ec.europa.eu/programmes/horizon2020/en/h2020-section/secure-societies-%E2%80%93-

protecting-freedom-and-security-europe-and-its-citizens, as well as Michel F. Bosco, "Trends on Security Research in Europe," Chapter 12 in this volume, 205–227.

26. "Surveillance: Ethical Issues, Legal Limitations, and Efficiency" (SURVEILLE), EU security research project, last accessed March 6, 2015, http://www.surveille.eu
27. "Detection Technologies, Terrorism, Ethics, and Human Rights" (DETECTER), EU security research project, last accessed March 6, 2015, http://www.detecter.eu
28. "Increasing Resilience in Surveillance Societies" (IRISS), EU security research project, last accessed March 6, 2015, http://irissproject.eu
29. "Rules, Expectations & Security through Privacy-Enhanced Convenient Technologies" (RESPECT), EU security research project, last accessed March 6, 2015, http://www.respectproject.eu
30. See http://www.surveilleadvisoryservice.eu, last accessed March 5, 2015.
31. Societal Impact Expert Working Group, "Report of the Societal Impact Expert Working Group," 11.

Part IV

Aspects of Programming for Homeland and Civil Security

11

Homeland Security within a School of Public Affairs

STEVEN A. PETERSON

Introduction

In February 2004, a set of faculty from across five disciplines (political science, public policy, public administration, criminal justice, and sociology) at Penn State Harrisburg developed a concept paper for active programming in homeland security. Some excerpts from that document:

Concept Paper
Center for Homeland Security and Terrorism Studies
Penn State Harrisburg
February 2004

> [...] We propose a Center housed in the School of Public Affairs at Penn State Harrisburg that has as its mission research and enhanced public awareness of terrorism and homeland security.
>
> The School of Public Affairs at Penn State Harrisburg is well-suited to be the sponsor of such a Center. It is located near the capital of one of the largest states in the nation, Pennsylvania, and is only 120 miles from the nation's capital. The Central Pennsylvania region is home to a number of military and defense installations and is a critical point in the major east/west and north/south transportation arteries in the eastern United States. Its long-established relationship with the U.S. Army War College (AWC) at Carlisle, 23 miles from its Middletown campus, provides an opportunity to work with military personnel who share an interest in the subject. With an interdisciplinary faculty comprised of specialists in [Political Science], Public Policy, Public Administration, Health Care Administration, and Criminal Justice, it can bring a variety of perspectives to bear on issues of homeland security and terrorism. [...]

> Three types of activities are envisioned to be sponsored by the Center. The first is basic and applied research on the subjects of terrorism and homeland security. Several faculty members are active in research and writing on the subject, and several others are engaged in research on related subjects (examples include interlocal cooperation, a key element in first-responder capability; policing issues; the political dynamics of nations and regions in which terrorism is a major issue, etc.). [...]
>
> A second major activity is public education and outreach. The Center will draw upon faculty, outside experts, and community leaders to develop a program of public education on topics of importance. [...]
>
> A third initiative is the preparation of trained professionals in the field of homeland security and anti-terrorism. Working through degree programs in Public Administration, Public Policy, and Criminal Justice, the Center will stimulate course development and internships to enable students to pursue careers and roles in organizations involved in homeland security and/or continue studies on subjects of importance to the field.

With this vision in mind, the School began a process of program development, centering on homeland security. This chapter focuses on several aspects of this: (1) The relevance of homeland security within a School of Public Affairs with a political science perspective included; (2) a brief history of the process of program development; (3) where the program is at today.

Homeland Security, Public Affairs, and Political Science: Substantive Linkages

Some elements of the study of homeland security are a bit afield from political science, such as cybersecurity, GIS and intelligence, public health preparedness, and agricultural biosecurity. Others, however, come from within a political science orbit—or, even more broadly, a public affairs regime.

One approach to defining the span of homeland security as a concept has noted some common themes across existing definitions:[1]

- The homeland security enterprise encompasses a federal, state, local, territorial, and tribal, as well as government and private sector approach that requires coordination;
- Homeland security can involve securing against and responding to both hazard-specific and all-hazards threats; and

- Homeland security activities do not imply total protection or complete threat reduction.

The School of Public Affairs has addressed a number of issues related to these concerns. Again, recall that the concept paper noted that we would be involved in teaching, research, and service/outreach in the domain of homeland security.

The Penn State Harrisburg School of Public Affairs has had political science and criminal justice faculty carry out research on terrorism.[2] They have developed and applied methods of study that are within the political science tradition.[3] Coursework in our online homeland security program focuses on policy and administrative issues associated with homeland security. Service and outreach? We worked to develop an interdisciplinary program, in concert with a regional non-profit organization, on homeland security in 2005; we have sponsored public speeches by former Pennsylvania Governor Thomas Ridge and one of his top associates in the then newly developed Department of Homeland Security (Admiral James Loy, who then became the first Administrator of the Transportation Security Administration, or TSA); we have presented seminars on use of "drones" in homeland security, as well as organized a large conference on the 35th anniversary of the Three Mile Island (next door to our campus) nuclear accident and evolving emergency management crisis in 1979 (thus, demonstrating our commitment to an all-hazards approach to homeland security).[4] We developed symposia on intelligence and homeland security that explored the policy and administrative elements of the homeland security enterprise. We have partnered with the Army War College on table top exercises for students that focus on the relevant political and administrative aspects of cybersecurity. Thus, homeland security has become an integral part of the School's mission of teaching, research, and service/outreach.

History of Homeland Security Programming in the School of Public Affairs: A Chronology

The first initiative was the concept paper[5] already described. This was not just rhetoric. Over the next few years, the School of Public Affairs engaged with the subject of homeland security. Next, we provide a brief chronology to suggest how the rhetoric became reality.

One benchmark: In the mid-2000s, the Penn State College of Medicine began an online master's degree program in health care and homeland security. Faculty within the School of Public Affairs began discussing the possibility of a policy and administration concentration within that program. Discussions were held with the Chair of the College of Medicine program. One scenario that was discussed: A concentration consisting of courses in homeland security administration and policy, terrorism and counterterrorism, policy analysis (a Master of Public Administration [MPA] course already being offered), and intergovernmental relations (another MPA course being offered). By fall 2005, discussion was started with Penn State Harrisburg officials, with special consideration of the Associate Dean of Graduate Studies. In the spring of 2006, a more interdisciplinary curricular model emerged, including a partnership with psychology (disaster psychology), a new public administration course in emergency management, a homeland security program-sponsored course in homeland security law, and a criminal justice course in terrorism and counterterrorism.

In 2006, the School of Public Affairs began involvement in a University-wide "Homeland Security Initiative," with a key component—the Homeland Security Coordination Council—and became a player within the larger enterprise. Among the members of this ad hoc group were the coordinator of the online master's program from the College of Medicine just mentioned above, one of Penn State University's lobbyists, a representative of the College of Agricultural Sciences Emergency Preparedness and Responsiveness office, the Dean of the College of Information Systems and Technology, Associate Director for Academic Programs from the World Campus (the distance delivery system for Penn State's online programming), the Associate Director of the Office of Military and Security Programs, the Chief Privacy Officer, the Associate Vice President for Research, the Director of the Justice and Safety Institute, the Director of the School of Public Affairs at Penn State Harrisburg, and so on. The group obviously included a wide range of individuals from many different offices.

The focus of this group was to serve as an information conduit as to which players were doing what with relationship to homeland security. The need for coordination was apparent from early on in the process. One person mentioned that it would be important to consider doing research on preventing incursions by hostile intruders via the shores of lakes, rivers, and oceans. Another member of the Council mentioned that Penn State Erie was already carrying out activities along these lines. The lack of knowledge as to what was being done in homeland security across Penn State made it clear that the Coordination Council was badly needed.

In 2008, after another successful partnership with a regional non-profit organization in presenting a symposium on homeland security, the School launched a website on homeland security. At about the same time, the Naval Postgraduate

School in Monterey, California, began advocacy for offering a postbaccalaureate certificate program (with four of their courses as the centerpiece). Those courses included the following from the extant graduate curriculum in homeland security: an introduction to homeland security, a course on multidisciplinary approaches to homeland security, another on strategic planning in homeland security, and a final course on knowledge into practice. Each course had a clear relationship to a School of Public Affairs.

The outline of the certificate had been discussed for some time. A committee composed of representatives from fifteen actors, including Texas A&M, the U.S. Army War College, United States Northern Command, and Penn State discussed development of a certificate program (the document was made available in June 2006). Conversation by the Penn State Coordination Council on the development of more comprehensive graduate programming led to promulgation of a document in June 2007. Here, discussion suggested several potential models. Three options for program delivery were spoken of—fully online, hybrid, and resident instruction. Over time, the fully online option was realized. Some preferred a full master's program with several specializations; others spoke of the desirability of at least beginning with more modest certificate programs.

As this was developing, the Naval Postgraduate School certificate program was offered to Penn State for possible online delivery. Two colleges at University Park declined the invitation. Then, the School of Public Affairs was offered this opportunity and accepted development and delivery of the program. The audience was National Guard and first responders. The School, in concert with the online delivery system, Penn State's World Campus, developed four courses and launched them in fall 2009. While these courses did not enroll well, the University, at the same time, under the leadership of the head of the College of Medicine's master's program in public health and homeland security, began formal, focused discussion of a Penn State multi-college online master's program in homeland security.

Given that the School of Public Affairs was already beginning to offer a certificate, it was given the possibility of creating one of five options in this program (other partners being the Colleges of Earth and Mineral Sciences, Agricultural Sciences, Information Sciences and Technology, and the College of Medicine). The School of Public Affairs and the College of the Liberal Arts were to develop and offer three core courses to give all students a common understanding of homeland security as a discipline. Against all odds, the collaboration of six colleges worked.

In 2009, the University also submitted a proposal to the Intelligence Community Center for Academic Excellence for a grant to support an effort to recruit intelligence professionals. The University saw this as an opportunity to wed intelligence concerns with homeland security. The School of Public Affairs became an actor within this process. We were assigned two primary responsibilities: organizing and delivering a symposium each fall and trying to organize student interest in possible intelligence careers. The primary focus, though, was on the symposia. The key organizer involved national figures in these symposia—including past heads of the National Security enterprise, officials from Homeland Security, and so on. Again, the focus was policy and administration, subjects well within the realm of political science and public affairs more generally.

In 2011, the School of Public Affairs created an Advisory Council for its Option in the Homeland Security master's program, with members having backgrounds from DHS (including Admiral James Loy as Chair of the Advisory Council), the Army War College, Ridge Global (consulting), General Dynamics Information Technology, Director of the Pennsylvania Department of Homeland Security, and the Pennsylvania Attorney General's Department, among others. By 2013, this Council had become the nucleus of a wider advisory group for the entire six-college collaborative effort, with each college involved contributing members to the Council, beyond the original Penn State Harrisburg group.

Where We Are Today

The above provides a narrative regarding why homeland security programming can fit naturally within a School of Public Affairs, including political science, and how the School shepherded the development of the effort forward. What about a preliminary evaluation of the initiative?

First, the School has become the administrative and academic home of the six-college online master's program. The first program chair was the head of the College of Medicine program as it evolved into a six-college effort. When he left Penn State, an interim chair was designated for a short period of time. Then, Jeremy Plant of the School of Public Affairs at Penn State Harrisburg was designated to serve as interim chair of the program. At this time, the Graduate School recognized the School as the program's administrative and academic home. After serving a year and a half, Plant (a political scientist who later served as coordinator of Penn State Harrisburg's MPA program) stepped aside as Alexander Siedschlag—another political scientist—took the reins of the program. It is clear that the original concept paper's statement that we ought to develop an educational mission has borne fruit.

Second, the Advisory Council of the School's program has become the nucleus of the six-college collaboration with respect to online education. Thus, the School's original initiative on this front helped to mobilize a larger effort.

Third, the desire to develop a research presence in homeland security within the school has developed. Several faculty at Penn State Harrisburg—including faculty across several schools with an interest in homeland security—have developed research initiatives in homeland security.

Finally, with regard to the goal of service and outreach, symposia, panel discussions, tabletop exercises, and revisiting events such as the Three Mile Island symposium have begun to create a presence in the local region.

Quo Vadis?

Thus far, the School of Public Affairs appears to have done well at advancing the hopes expressed in the 2004 concept paper. What about the future?

First, homeland security as an issue is not unique to the United States. An interest in the general subject, however labeled, is manifest in other countries. One of the tasks facing the School and the overall degree program is to develop a greater comparative and global perspective. For instance, Alexander Siedschlag has spearheaded development of a new course on comparative homeland security as well as a program concentration in international homeland security.

Second, homeland security has become an element of graduate and undergraduate students' education. Students in our online Master of Public Administration program not uncommonly sign up for homeland security online classes to create an ad hoc concentration in homeland security. We have developed an online undergraduate minor in homeland security, and majors in online programs in criminal justice and political science commonly include these courses in their schedule—and sign up as minors. Thus, we see another linkage between public affairs and homeland security.

Third, we have had a fund endowed to support outreach to the larger community in cybersecurity. We are seeking funding for a Center for Homeland Security, to facilitate research support and outreach to the larger community. How successful this will be is still an open question. But, if successful, Penn State Harrisburg will have a larger footprint within the region.

Thus, in the final analysis, homeland security has become a central part of the activity within the School of Public Affairs at Penn State Harrisburg—with political science and public administration as key partners in this effort.

Notes

An earlier version of this work was prepared for presentation at a short course on Homeland Security organized by Penn State's Intercollege Master of Professional Studies in Homeland Security Program at the Annual Meeting of the American Political Science Association (APSA), Washington, D.C., August 2014.

1. Shaun Reese, "Defining Homeland Security: Analysis and Congressional Considerations," Congressional Research Service, CRS Report for Congress, R42462, 2013, 9, http://fas.org/sgp/crs/homesec/R42462.pdf; see also Robert W. Smith, "What Is Homeland Security?" *Journal of Public Affairs Education* 11, no. 3 (2005); Christopher Bellavita and Ellen Gordon, "Changing Homeland Security: Teaching the Core," *Homeland Security Affairs* 2, no. 1 (2006), http://www.hsaj.org/hsa/volII/iss1/art1
2. Jennifer Gibbs, "The Relationship between Foreign Military Presence and Terrorist Attacks on Police" (paper presented at Academy of Criminal Justice Sciences, Philadelphia, PA, 2014); Michael Kenney, *From Pablo to Osama* (University Park, PA: Penn State University Press, 2008).
3. Robert A. Pape, *Dying to Win* (New York: Random House, 2005); Mia Bloom, *Dying to Kill: The Allure of Suicide Terror* (New York: Columbia University Press, 2007).
4. See "TMI@35: The Three Mile Island Nuclear Crisis in Perspective. Analyses, Stories, Policies. Documentation of a Conference Held at Penn State Harrisburg, March 27–28, 2014," Penn State Homeland Security Programs, last accessed March 29, 2015, http://sites.psu.edu/tmi35
5. Jeremy Plant and Michael Kenney, "Center for Homeland Security and Terrorism Studies," concept paper, Penn State Harrisburg, School of Public Affairs, 2004.

12

Trends on Security Research in Europe

MICHEL F. BOSCO

Security Research in Europe—Recent Roots

Inspired by Events That Occurred in the USA

Security research in Europe developed as a reaction to the September 11, 2001 terrorist attacks in the United States of America.[1] These events had an electroshock effect on the U.S. Administration, triggering the deepest ever restructuring of federal institutions dealing with security and the birth of the Department of Homeland Security. They also raised consciousness in Europe about the limited capacity of the nations to adapt to threat and to innovate in their response to malfunctions or intentional disruptions in their societies.[2] At that time, only a handful of European Union (EU) Member States had defense-related research going on that could serve a dual purpose, but where the innovative impact on civil security was a spin-off effect of novel technologies or processes resulting from military requirements. In 2003, the European Commission, the executive branch of the European Union, decided to explore the feasibility of financing security research at EU level, and expressed its intention to enhance the EU's scientific and technological capabilities for ensuring the security of European citizens.

The initiative was clearly proposed as a political response to an issue under the spotlight: the September 11, 2001 terrorist attacks on the USA, a target to which several EU Member States felt closely associated—and thus also at risk. It was also an attempt by governments to address the impact and resilience of the European economies to the natural and man-made disasters that occurred in Europe in 2002. European nations had proved unprepared to react in a coordinated and "modern" way to both a devastating flooding in August (that affected a large area engulfing Austria, Bulgaria, the Czech Republic, and in particular the Bohemia region, Eastern and Southern Germany, Hungary, northern and central Italy, Romania, Slovakia, the north east of Spain, and the United Kingdom) and the "Prestige" oil spill off the coast of Galicia, caused by the sinking of an oil tanker (that impacted France, Portugal, and Spain). Public opinion polls at that time showed that European citizens expected government to take a more international approach when it comes to their security. The European Parliament was encouraging the Commission to undertake actions that would strengthen Europe's long-term security capabilities.

Tackled from the Viewpoint of Industrial Policy

The initiative was also an attempt by the European Union to move closer to playing a role in European defense—so far a domain exclusively within the competences of individual nations—by introducing industry policy measures. A press release from the European Commission makes it explicit that "Europe was paying a very high price for the artificial, and uniquely European, separation between civil and military research."[3] It further states "new and more sophisticated security threats, combined with an increasing aspiration from citizens for security was requiring Europe to adopt a more structured and European approach to security research."[4] "Fostering co-operation between the traditionally distinct [civil and military research] sectors, by creating an environment in which European defense-related companies could give better value for money" was presented as the proper way forward.[5]

The EU was being depicted as a place where the fragmentation and compartmentalization of defense research was preventing its industry from investing in certain key technologies.[6] The European Parliament drew attention to some of these issues in its resolution on defense industries of May 13, 2002.[7] In contrast, the USA was being quoted as an example of an economy where defense-related research plays a major role in innovation, which benefits the whole industry including the civilian sector in terms of market access and costs.[8] Such a statement was correct, but built in part on a half-truth: The EU was indeed fragmented into 15 different markets in the defense sector, but in the largest of these markets (such as France, United Kingdom, Netherlands, or Germany) synergies existed with the civil security sector.[9] By associating civil security with defense, the European Union had found a way to pull the momentum established in support for greater European integration in the field of civil security, behind greater European integration in defense.[10]

Further Anchored following the London and Madrid Bombings

Through its March 2003 Communication: "Towards an EU Defence Equipment Policy,"[11] which called for the establishment of a European defense equipment market, the European Commission invited EU Member States, industry, and the scientific community to join forces to identify common needs and pool know-how and investment to develop technologies critical for Europe's security. A "Group of Personalities"[12] helped to establish the vision and aims for European civil security research, and the European Commission mobilized—a first at European level—€65 million to define and foster the development of a long-term strategy, and form the basis of a future, fully-fledged EU civil security research program. In spring 2004, this group concluded:

> Europe has high quality research institutes and a substantial and diverse industrial base from which to address technology requirements in the security domain. However, structural deficiencies at the institutional and political level hinder Europe in the exploitation of its scientific, technological and industrial strength. The dividing line between defence and civil research; the absence of specific frameworks for security research at the EU level; the limited cooperation between Member States and the lack of coordination among national and European efforts all serve to exacerbate the lack of public research funding and present major obstacles to delivering cost-effective solutions.[13]

The March 2004 bombings in Madrid, and the July 2005 attacks in London came up in a sad, timely manner to amplify across Europe a feeling of empathy: The "We are all Americans," proclaimed in France's *Le Monde* newspaper on September 12, 2001 had evolved into "We are all Europeans." In March 2004, the European Commission action paper in response to the terrorist attacks on Madrid was already proposing to increase the European effort against terrorism:

> We need to urgently review whether we have adequate measures in place to allow us to monitor and trace bomb-making materials such as explosives, detonators, and radioactive sources as well as precursors. Detection and traceability should become our key words here. The tracing and checking of dangerous goods and explosives should be made possible by creating new databases or upgrading existing databases such as SIS II [the Schengen Information System] with new functionalities, as well as making full use of advanced technologies such as satellite-enhanced (GALILEO) RFID (Radio Frequency Identification Device) tracking. All this action needs to be underpinned by a re-prioritization of our research resources by fostering their increasing dedication to security matters,

in particular inter-operability. The present amount of €65 million devoted to security-related research is woefully inadequate.[14]

Taking Account of the Shift in Public Perception of Liberties vs. Security…

The trend was clear, but the road ahead remained bumpy. In spite of the notable shift in public perception across Europe, as a reaction to the series of terrorist attacks, that security and security research in particular may require action at EU level, civil liberties watchdogs ensured that a debate would take place about the appropriateness of such developments.[15] They highlighted that out of the 27-strong "Group of Personalities," who had been laying down the first principles for European security research, eight were CEOs or senior directors of multinational corporations, including Europe's four largest arms manufacturers. They argued that the various structures that subsequently advised the European Commission counted few participants from communities of civil users of security research, though they included some representatives from civil protection organizations and ministries of interior, and even fewer members of civil liberties or privacy organizations. They also regretted that the Commission issued no green paper on security research for setting out possible policy options prior to launching a fully-fledged EU civil security research program in 2007. The debate continues today, and the balance to strike between liberties and privacy versus security remains a central issue.

…But with Little Impact on Synergies between Military and Civilian Security Research

Currently, security research undertaken at EU level is funded as part of the Horizon 2020 European Framework Program, a pool of almost €80 billion dedicated to research, development, and innovation in Europe. Its rules, co-decided by the European Parliament and the Council, are quite clear about the civil security nature of the research it may finance.[16] They stipulate, as a matter of principle to be applied to the whole of the Horizon 2020 Framework Program, that research shall have an exclusive focus on civilian applications:

Article 19—Ethical principles

1. All the research and innovation activities carried out under Horizon 2020 shall comply with ethical principles and relevant national, Union and international legislation, including the Charter of Fundamental Rights of the European Union and the European Convention on Human Rights and its Supplementary Protocols.
   Particular attention shall be paid to the principle of proportionality, the right to privacy, the right to the protection of personal data, the right to the physical and mental integrity of a person, the right to non-discrimination and the need to ensure high levels of human health protection.
2. Research and innovation activities carried out under Horizon 2020 shall have an exclusive focus on civil applications.[17]

In the description of research areas to be covered to enhance "secure societies," however, they do refer to:

7. Secure Societies—Protecting Freedom and Security of Europe and Its Citizens
[…]
7.3. Broad lines of activities
[…]
(h) support the Union's external security policies, including conflict prevention and peace-building.[18]

The text includes no explicit reference to "dual-use" research, and in subsection "7.2. Rationale and Union added value," no reference is made about the "dividing line between defence and civil research" which, in previous years, had been identified by the European Commission[19] and by the European Parliament[20] as one of the major "structural deficiencies" of security research and industry in Europe. It is only in the subsequent annual work-programs, which give detail about the research topics to be addressed, that "dual-use" is used explicitly in the case of research on chemical, biological, radiological, nuclear, and explosive (CBRN-E) detection: "Explosive detection (CBRN): Dual-use aspects will be considered with possible synergies being established with the European Defence Agency (EDA)," with the argument that "higher cost-efficiency through dual-use applications" would give greater impact to research in CBRN-E.[21]

From National to Union-wide User of Security Research Communities in Europe

No Unique Model among EU Member States…

The ways in which communities of practitioners and end-users of security research are organized in Europe continue to vary considerably from one country to another.[22]

The split between civilian and military responsibilities in the control or the implementation of security-related missions differs. Some EU Member States feature only civilian police. In other Member States, police forces with military status exist alongside civilian police. Even in such cases, the chains of command may be different. In Romania, for instance, the Jandarmeria, whose staff has a military status, acts under the authority of the Minister of Interior in peacetime; in France, the Gendarmerie acts under the dual authority of both the Ministry of Interior and the Ministry of Defense; in Spain, the Guardia Civil is under the authority of four Ministries. This variety of systems quite naturally results from the varying histories of these nations. Changes continue to occur: In Italy, for instance, the Carabinieri, which once constituted a police force under the Ministry of Defense, are now under the authority of the Ministry of Interior. In spite of their different chains of commands, these four police forces are—together with similar forces from two more Member States (the Netherlands and Portugal)—engaged in joint operation under the European Gendarmerie Force (EGF).[23]

Similar variations apply to Coast Guards: a unit of the police in Cyprus, but a separate organization in Greece; a purely civilian entity in Germany, but a unit of the Navy in Norway—not an EU nation, but applying the *Schengen acquis.*[24] The responsibility for carrying out coast functions is distributed across 316 public authorities in the maritime Member States of the European Union.[25] Many of these authorities are participating in operations coordinated by FRONTEX (see below) and undertaking joint activities with peer organizations in other countries through a total of 70 collaboration structures operating at the international, EU, and regional levels. There are 41 bilateral and multilateral cross-border cooperation agreements between EU Member States, some of which involve also non-EU countries.

Firefighting organizations are national in Italy and in a few other Member States, but regional in Germany's *"Länder"* (federal states). In France, they are part of the Navy in Marseille, of the Army in Paris, but county or city civil servants elsewhere. In Belgium, firefighters are under the authority of the regions, but civil protection at large is a service of the Federal Ministry of Interior. Different cultural backgrounds also contribute to the heterogeneity among such organizations:[26] As regards to gender, for instance, the Estonian police force was composed of almost 40% of female officers in 2008, versus only 8% in the Spanish police. Such diversity among communities of security practitioners in the 28 EU Member States makes operational coordination complicated, an issue that is being tackled in a quite successful way through the establishment of agencies at European Union level (as addressed below).

...But Joint, Legal Responsibilities Exist for the EU further to the Abolition of Internal Borders for Goods and People

Major legal steps were taken, however, making operational coordination mandatory and coordinated planning necessary: the European Union Customs Union; the Schengen cooperation for the abolition of border control at internal borders; the decision to establish the Union Civil Protection Mechanism; and others.

The Customs Union

The Customs Union abolishes tariffs within the internal borders of its members, while it establishes common tariffs at its external borders. De facto, the vast majority of customs checks of goods arriving in the EU are taking place at airports, at its land borders with Eastern European countries, and at ports on the coast along the Atlantic or the Mediterranean, the Black, the North or the Baltic Sea. Unlike in the U.S., where customs are the responsibility of a federal agency, the external border of the EU is under the distributed protection of the Member States, each for its sections of the common border. Pan-European cooperation is established for matters that can be dealt with in a more central manner, including through the Customs Laboratories European Network for scientific expertise. [27]

The Schengen Agreement

The Schengen Agreement initiated the establishment of a territory within Europe, where passports are not needed and border control is abolished. This definitely encourages the free movement of goods, information, money, and people within the area. As a prerequisite, however, the signatories of the Schengen Agreement agreed to strengthen controls at their external border. Again, operations at those borders are not implemented by any federal agency such as in the U.S., but rather by national bodies. One step further toward greater integration has occurred with the setting-up of the "European Agency for the Management of Operational Cooperation at the External Borders of the Member States of the European Union," or FRONTEX.[28]

Even as the paragraphs above depict the slow evolution from a very fragmented situation toward a more coordinated and coherent situation, the picture remains complex. Only 22 of the 28 EU Member States participate fully in the Schengen area (the UK and Ireland having opted to stay out, and four recent Member States having not yet completed the Schengen accession process). Conversely, not all Schengen countries (e.g., Switzerland) are EU Member States, and thus not part of the Customs Union, so some controls on goods entering or leaving the Customs Union are unavoidable.

The Union Civil Protection Mechanism

The Union Civil Protection Mechanism[29] is to support, coordinate, and supplement the actions of the Member States in the field of civil protection in improving the effectiveness of systems for preventing, preparing for, and responding to natural and man-made disasters of all kinds within and outside the Union. It promotes the sharing of resources among Member States, and provides, at EU level, for the establishment of information systems, risk analysis and risk-management capabilities, and the processes necessary to identifying and filling gaps in the response capacities, and to facilitating the operational coordination of national means.

FRONTEX

The "European Agency for the Management of Operational Cooperation at the External Borders of the Member States of the European Union" (FRONTEX) is mandated to coordinate cooperation among border authorities, as they exist in the EU Member States, including through the planning, the operational coordination, the implementation, and the evaluation of specific operations. FRONTEX deploys in a permanent manner several European Border Guard Teams constituted of officers from several national border authorities who are experts in different areas of border management, including land and sea border surveillance, dog handling, identification of false documents, and in the establishment of nationalities of irregular migrants. For instance, in 2014, FRONTEX launched a Joint Operation called "Triton"[30] to face flows of irregular migrants from Africa and from the Middle East in the Mediterranean Sea, to cover the south coast of Italy and Malta. In this Joint Operation Triton, supported by 21 European countries, FRONTEX teams as well as sea- and airborne means cooperate with Italian border authorities: the "Guardia di Finanza" and the Coast Guard.

FRONTEX must also perform a number of tasks to the common benefit of all border authorities in the EU: It maintains an extensive database of equipment available in Europe to be deployed in joint operations; it develops training standards, with a view to facilitate joint operations among authorities from several nations and to promote a common application of the *Schengen acquis*; and it provides common information on the ongoing situation at the external borders of the EU, generating tactical intelligence usable for operations as well as a more strategic situational awareness about emerging risks and the current state of affairs at the external borders.

FRONTEX also collects, analyzes, and validates with border authorities and border-control personnel across Europe their needs in regards to mid- or long-term innovation. FRONTEX, to that effect, builds on its closeness to actual border-control personnel and organizations, and state-of-the-art research to create innovative prototypes and new technologies, essentially within R&D (research and development) projects financed by the European Commission, and in a smaller measure in the national R&D programs of a few EU Member States. As for Europe's southern maritime border, for instance, only few authorities performing coast guard functions engage in research and development activities, or are aware of research undertaken in other, neighboring Member States. EU-level funding in research related to maritime security is significant. FRONTEX helps raise awareness of past and existing research that could bring innovation in coast guard functions, and is in a position to advise the European Commission when it establishes roadmaps for its R&D spending.

EUROPOL

EUROPOL ("European Police Office") is the European Union's law enforcement agency that supports and assists its Member States in preventing and combating all forms of serious international crime and terrorism.[31] EUROPOL does not have operational capacity of its own, but rather coordinates that of its EU Member States, in particular through the development of information capabilities. EUROPOL aims at becoming the central information hub for policing in the EU and of specialist expertise that law enforcement authorities in the Member States can rely on. It carries out a number of high-profile cross-border investigations—up to 20,000 each year. In 2014, for instance, it ran operation Archimedes,[32] the biggest-ever coordinated crackdown on organized crime in Europe to date making more than 1,000 arrests, including 170 in connection with people trafficking; checking 10,000 irregular migrants; saving 30 Romanian children from trafficking; and seizing 440lb of heroin, 1320lb of cocaine, and almost 2900lb of cannabis. Archimedes involved all 28 EU Member States plus police forces in six other countries: Australia, Colombia, Norway, Serbia, Switzerland, and the USA.

EUROPOL has established itself as the lead organization in Europe in the emerging area of cybersecurity. Since 2013, it hosts the European Cybercrime Centre (EC3), which cooperates not only with national law enforcement authorities in the EU but also with the U.S. Federal Bureau of Investigation (FBI) and U.S. Immigration and Customs Enforcement (ICE). EC3 supports the EU Member States and Institutions in building operational and analytical capacity for investigations and cooperation with international partners. It also runs or coordinates operations against cybercrime committed by organized groups to generate large criminal profits, which causes serious harm to the victim such as online child sexual exploitation, or which affects critical infrastructure and vital information systems in the European Union. In November 2014, EC3 coordinated the action of law enforcement and judicial agencies in 16 European countries with that of the FBI to bring down several marketplaces on the dark-net, including the infamous Silk Road 2.0.[33] The action resulted in stopping the sale of weapons, drugs, and murder-for-hire services, leading to 17 arrests of vendors; taking down over 410 hidden services; and seizing bitcoins worth approximately $1 million as well as large amounts of cash, drugs, gold, and silver.

eu-LISA

The "European Agency for the Operational Management of Large-scale IT Systems in the Area of Freedom, Security and Justice" (eu-LISA) hosts and maintains the European information systems necessary to manage asylum seekers' applications, visa information, and information on persons who may have been involved in a serious crime or who may not have the right to enter or to stay within the territories of the Schengen area.[34] The core task of this agency is to keep the systems functioning 24 hours a day, seven days a week, ensuring the continuous, uninterrupted exchange of data among national authorities. The agency also organizes training for expert users of the systems, and monitors research activities in areas that may lead to innovation in the management and, in particular, the security of information systems.

ERCC: The EU Emergency Response Coordination Centre

The EU Emergency Response Coordination Centre (ERCC) contributes to the implementation of the Union Civil Protection Mechanism, and facilitates solidarity among Member States when one is calling for help when facing a disaster.[35] It collects and analyses real-time information on disasters, monitors hazards, and prepares plans for the deployment of experts, teams, and equipment, including through the preparation, validation, and simulation of a set of typical disaster scenarios. The ERCC works with Member States to map available assets and coordinate the EU's disaster response efforts by matching offers of assistance to the needs of the disaster-struck country. Similarly to what FRONTEX does by interacting with border guards, the ERCC collects, analyzes, and validates with first responders across Europe what are their needs in terms of mid- or long-term innovation, as well as what can be drawn from state-of-the-art research and new technologies developed within R&D projects financed by the European Commission.

From National to EU-wide Security Research Programs

Only a Few EU Member States Are Concerned

Very few EU Member States were running R&D programs focused on civilian security applications[36] at the time of September 11, 2001. In 2004, on the eve of the launch of the first EU-wide Security R&D Program,[37] only seven[38] Member States had ongoing national investments in the field. Collectively, they were dedicating less than €130 million to such activities yearly. While the large Member States in the group (France, Germany, and the UK) are now addressing a broad cross-section of security research topics, the smaller ones favor a more focused approach, which may vary from one year to another. By the mid-2010s, national and EU-level programs continue to coexist, the combination of the national budgets summing up to about half of the corresponding EU budget.

From EU Policies to EU Research

With the emergence of European Union policies in the area of security as described above, EU research can now develop with a view to providing all of the Member States (and not just a few, as was the case in the past) with direct access to the most innovative and interoperable solutions, while taking advantage of the specialized research and industrial resources already established in some countries. For this to happen effectively requires the establishment of EU-compatible mechanisms throughout the R&D cycle—from the collection of user requirements to the actual procurement of solutions, including research programming, funding, implementation, and industrial uptake.

Starting with Users' Requirements

"Users," even in the very few EU Members States where a process has been established by which they can express their innovation needs to central administrations, have however little time to invest in such processes—their priority is operations—and innovation often continues to result from an industry push. The gathering of practitioners' perspectives remains even a more difficult issue to address at European level given the very fragmented user population, in terms of missions, control chains, procurement capabilities, languages spoken, and the scarcity of organizations investing in the analysis and the expression of such needs and requirements at pan-European level.

To tackle this situation, the EU research, development, and innovation program for secure societies[39] is financing in a more systematic manner networks of users' organizations—be they national, regional, or local—willing to get together and spend staff working time on analyzing, discussing, and expressing their requirements, while being advised about the actual technological possibilities and industrial capabilities. The specialized EU Agencies—FRONTEX, EUROPOL, etc.—are invited to participate in the work of such networks and even to lead in some areas. For instance, the European Network of Law Enforcement Technology Services (ENLETS)[40] was established to strengthen police activities and cooperation, promote the use of modern technologies in the process of exchanging information, knowledge, or experience. ENLETS collects user requirements across Member States and interacts with EUROPOL and INTERPOL. ENLETS was founded in 2008, and it develops steadily but slowly given the diversity within its membership.

The European Explosive Ordnance Disposal Network (EEODN), also founded in 2008, discusses preventive measures in fighting terrorism and exchanges the best worldwide experience in eliminating possible threats. Recently it has called for more research toward neutralizing "dirty bombs"—improvised explosive devices used to spread radiological and chemical substances. It also organizes pan-European testing of innovative equipment and ways of detecting and neutralizing explosives.

Supporting Competitiveness

To promote industrial competitiveness and innovation, the European Commission grants the vast majority of its financial support to projects undertaken by temporary consortia consisting in a mix of academic, industrial, and practitioners' organizations. The support provided by the European Commission rarely exceeds 50% of the actual cost of the activities performed within each project. A consortium gathers the competences necessary to perform the R&D work and prepare for its "exploitation," that is, the steps required to translate research results into innovation. The organizations that participate in a consortium do agree among themselves how they share pre-existing know-how and intellectual property rights (IPR), and the results of the activities co-financed project. The European Commission is also working toward furthering the pan-European certification of security-related products in some sectors, starting with the promotion of standardization and planning for EU-level regulations in well-motivated cases.

Addressing Industrial Issues Quite Specific to the Field of Security

Programming research work in the field of security needs to take account not only of the requirements expressed by the future users of technologies and of innovative products, but also of the industrial reality and the market environment. There are almost no pan-European norms applicable as regards to the performance and characteristics of security-related products. In addition, the very specialized characteristics of security-related products limit the size of their markets, which often belong to the "niche-market" category, with little incentive for industry to invest, and often with established monopolistic positions as a result—the latter contributing to the issue of disproportionate prices. Consequently, industry tends to limit their offers, their capabilities to design and to produce, and their business strategies are often constrained by their (public) customers' limited budgets. The European Commission's approach to the issue combines the use of both carrots and sticks.

When financing R&D in the lower technological readiness levels (TRL), which is quite upstream from market, the European Commission co-funds research projects, with participating industry sharing the costs—typically 50%—but keeping ownership on the IP generated within the projects.

When funding R&D at higher TRL, the Commission proceeds through pre-commercial procurement of research, as follows: first, by gathering a share of the funding required—typically 30%—from a consortium of public authorities in several EU Member States, potential buyers of the technological innovations that are expected to result from the projects (this to ensure that research will satisfy common, cross-border requirements); second, by giving these potential buyers the right to require participating industry—that is being reimbursed 100% of its costs while retaining IPR—to grant non-exclusive licenses to European third parties to exploit the results under fair and reasonable market conditions; third, by financing two industrial competitors running parallel research streams until the end of each project.

By conditioning its financial participation to the endorsement of fair conditions to non-participating Member States, the European Commission intends both to limit the risk of disproportionate prices, and to encourage the design of solutions allowing for easier sales across the EU. Such conditions are sometimes questioned by industry, but a study has shown that, generally speaking, under the assumption of a fully funded R&D-only contract, the EU legal provisions seem to be marginally more contractor-friendly than those of the U.S.[41] This result seems to be mainly due to the presence of stronger "march-in rights,"[42] stricter obligations to protect results, and generally broader licensing rights for the contracting authority in the U.S. legal regimes. The U.S. legal regime's IP provisions also have slightly stricter rules around the contractor getting title to inventions than in the EU, making it comparatively easier to lose title if certain formal deadlines are not met. In sum, the "tough" provisions of the U.S. legal regime seem to reflect the stronger IPR commercialization policy-push inherent in the Bayh-Dole Act[43] of 1980.

Addressing Transnational, Trust-Related Issues

"Europeanization" of any aspect of security research requires building or enlarging circles of trust: In Europe, as in most regions around the world, trust in security-related matters is first and foremost established in line with national sovereignty. In each country the government, the relevant public authorities, industry, and research centers agree to share information and know-how among themselves, with a view to ensuring innovation and performance in addressing security issues in general on a national basis—within what can be called "national silos of trust." In such matters, exchanges beyond those organizations within the silo are channeled, in general, through the government and the relevant public authorities.

"Europeanizing" trust as regards security matters requires, de facto, reaching beyond the existing national silos of trust and, for the European Union, as far as to involve its 28 Member States. The institutional setting as it exists within Europe allows for two main approaches toward addressing the issue: the establishment of intergovernmental treaties, such as the Schengen Agreement at its inception, and European Union regulations—that is, laws applicable within the 28 Member States, with specific, partly supranational (sovereign) roles associated with the European Union institutions. With the national silos as they exist, the intergovernmental approach is probably the easier to start with, but it fails gathering the 28 Member States in most cases—and in the case of the Schengen Agreement, even after its inclusion in the EU Treaties, since some Member States have opted out.

The EU R&D program in security is a program of the European Union, but each project that belongs to the program gathers organizations from a number of countries. This allows for reaching beyond the national silos of trust—on one specific, limited matter—and for expanding partnerships. The EU R&D program in security definitely contributes to "more Europe," but it has its limitations: projects are of a finite duration—typically two to five years—and are ruled by contractual arrangements, whose continued existence would require the establishment of corresponding EU regulations. In the very many cases when the circle does not encompass the 28 Member States, use of the EU mechanism for "enhanced cooperation"[44] may be envisaged. Enhanced cooperation allows those countries of the Union that wish to work more closely together on one specific matter to do so, while respecting the legal framework of the Union. The recourse to "enhanced cooperation" within the EU is, however, a matter of debate;[45] some consider it the renouncement to an actual Union. However, as envisaged in the European Union treaties, security issues could become subject to enhanced cooperation at some point in time in the future.

Increasing Synergies between Military and Civilian Security Research

The budget authorities of the European Commission, that is the Council and the European Parliament, are demanding for optimizing the use of taxpayers' money. As explained above, security research in the Horizon 2020 Framework Program must have an exclusive focus on civilian applications on the one hand, while it needs to increase synergies with military research.[46] To that effect, the European Commission is cooperating with the European Defence Agency (EDA). Among its main objectives, the EDA is to enhance the effectiveness of European defense research and technology,[47] in particular when addressing the Research & Technology needs of the Common Security and Defence Policy.[48] The European Commission and the EDA facilitate the interaction between their projects through the organization of joint workshops and the cross-participation of officers in the steering and advisory structures of such projects. The European Commission encourages civilian research projects to look into the outputs from projects launched by the EDA and occasionally supports the adaptation of such outputs to civilian requirements. The EDA acts in a reciprocal manner, which results in enabling potential markets for innovative, dual-use solutions.

Under Citizens' Watch

The debate that started in the mid-2000s at the inception of the first EU security research initiative is obviously continuing, since the central concern remains: What is the right balance between liberties and privacy versus security? The European Commission is addressing such concern in various ways.

First, many civil rights and watchdog organizations[49] continue to monitor and to produce reports on the management by the European Commission of the EU R&D program in security and on the individual projects that it finances through this program. Such reports are usually subject to public debates, including in the media.

Second, the European Parliament is continuously monitoring the way the European Commission manages the program. Members of the "Civil Liberties, Justice and Home Affairs" (LIBE) committee of the European Parliament have asked tens of written questions since the beginning of the 2010s. The concerns that they express have been focusing on a few but acute issues: Research in security may lead to restriction of civil and human rights, for instance when it addresses enhanced surveillance and facial identification on video streams, the use of drones, and tracking systems. Research in security may also generate know-how of dual use, that is civilian and military, which requires reviewing the work of the researchers from an ethical perspective.

To ensure that the EU R&D program on security complies with its regulations, which cover such matters, the European Commission has set up mechanisms to assess the compliance of each project proposal with ethical principles,[50] including privacy and the respect of civil rights;[51] with the exclusive focus of the proposed activities on civilian applications; and with export control (and non-proliferation) rules.[52] The European Commission also assesses the need for some projects to classify certain project deliverables[53] to protect Europe against the malevolent use of research results. Furthermore, the European Commission has requested the European standardization organizations to develop a standard, modeled on existing quality management schemes, but applied to the privacy management in the design, development, production, and service provision processes of security technologies and services. Such a standard aims at allowing manufacturers and service providers to develop, implement, and execute a widely recognized "privacy-by-design"[54] approach in their processes.

For a Mature EU-wide Security Research Program by 2020

With its budget adopted until 2020, the EU program for security R&D managed by the European Commission is now taking shape, taking into account the larger variety of constraints described above. Its success for the years 2014–2020 will indeed be measured against its capacity to address more European requirements in an integrated way, when for many years such requirements used to consist essentially in the collated national interests in the Member States with limited EU added value. In its current format, the work program proposed by the European Commission is aligning the strands of research in the area of security to target long-term—and sometimes very long-term, even utopian—and all-hazards objectives, in the following areas:

- Border security and immigration:
    - For maritime security and immigration: Establish an integrated, distributed European system that builds upon the capacities of the individual EU Members States, in particular in the Mediterranean Sea. Innovation gaps include the small boat long-range detection, and 3D situational awareness.
    - For border crossing points: Develop smart border systems functioning without gates, while preserving privacy and other individual rights.
    - For land border and surveillance: Equip European border with advanced systems at affordable prices, including with under-foliage detection capabilities.
    - For supply chain security: Integrate the various, promising, innovative solutions developed throughout EU projects during the past five to ten years into robust and affordable systems.

- External security of the EU:
    - Develop an integrated information system to support the actions implemented by the EU under its Common Security and Defence Policy (CSDP).
    - Dramatically shorten the reaction time between the occurrence of an event and the subsequent EU decision for an EU CSDP intervention by improving interoperability among national and EU control centers and other capabilities.

- Fight against crime and terrorism:
  - Develop a common EU mobile forensics laboratory integrating the various tools and techniques developed, building upon EU research projects undertaken during the past five to ten years.
  - Develop modular, ubiquitous detectors for security-related applications.
  - Establish a common European approach to policing based on a common EU technology platform.

- Disaster resilience:
  - Develop a common EU approach to crisis management, also based on a common EU toolkit of techniques and processes addressing prevention, detection, protection, and intervention.
  - Establish a common European communication platform addressing first responders to first responder communication, as well as communication from authorities to the public.
  - Develop domain-specific but comprehensive protection of elements of the EU critical infrastructure, addressing both physical and cyber threats.

- Digital security and privacy:
  - Demonstrate the viability and maturity of state-of-the-art security, privacy, and trust solutions in real life environments, and involving end-users.

In each of these areas, projects are being supported with an overall annual budget approximating €200 million, with the dual objective to improve European competiveness and help make Europe a more secure society.

Notes

1. "Since then [September 11, 2001], specifically among countries of the Western hemisphere, traditional security paradigms and theories have been critically questioned and the different security cultures and perceptions have resulted in diverse security and defence policies as well as in security research efforts of individual countries," Khoen Liem, Daniel Hiller, and Christoph Castex, "A European Perspective on Security Research," in *European Perspectives on Security Research,* ed. Klaus Thoma (Berlin and Heidelberg: Springer, 2011), 13, http://www.acatech.de/fileadmin/user_upload/

Baumstruktur_nach_Website/Acatech/root/de/Publikationen/Englisch/acatech_diskutiert_European-Research_WEB.pdf

2. "A Secure Europe in a Better World. European Security Strategy," Brussels, December 12, 2003, http://www.consilium.europa.eu/uedocs/cmsUpload/78367.pdf
3. European Commission, "European Industry Leaders and EU Policymakers Meet to Plan for Security Research," press release, IP/03/1351, Brussels, October 7, 2003, http://europa.eu/rapid/press-release_IP-03-1351_en.htm
4. Ibid.
5. European Commission, "More Research for Europe—Towards 3% of GDP," Commission Staff Working Paper in Support to the Commission Communication, COM(2002)499, September 11, 2002, http://ec.europa.eu/research/era/pdf/com 3percent_en.pdf
6. Thomas Banchoff, "Institutions, Inertia and European Union Research Policy," *Journal of Common Market Studies* 40 (2002).
7. Carine Piaguet and Frédéric Gouardères, "Fact Sheets on the European Union—Defence Industry," European Parliament, March 2014, http://www.europarl.europa.eu/aboutparliament/en/displayFtu.html?ftuId=FTU_5.9.5.html
8. European Commission, "More Research for Europe."
9. E. Anders Eriksson, Mattias Johansson, and Martin Lundmark, *Civil-Military Synergies in Research and Technology: Comparison of Sweden to Six European Nations* (Stockholm: FOI, 2014).
10. Anecdotally, the same, factual root—the September 11 events—was used as a rationale for bringing together defense and civil security in Europe, while in the USA it led to creating a new Department of Homeland Security that consolidated security-related activities outside of the scope of defense.
11. European Commission, "Communication from the Commission to the Council, the European Parliament, the European Economic and Social Committee and the Committee of the Regions: European Defence—Industrial and Market Issues. Towards an EU Defence Equipment Policy," COM(2003) 113(01), Brussels, March 11, 2003.
12. A group of 27 senior "wise persons" with interests in security and security-related research, comprised of politicians, industrialists, and researchers.
13. *Research for a Secure Europe Report of the Group of Personalities in the Field of Security Research* (Luxembourg: Office for Official Publications of the European Communities, 2004), http://ec.europa.eu/enterprise/policies/security/files/doc/gop_en.pdf
14. European Commission, "European Commission Action Paper in Response to the Terrorist Attacks on Madrid," MEMO/04/66, March 18, 2004, http://europa.eu/rapid/press-release_MEMO-04-66_en.htm?locale=en
15. See for example Leigh Phillips, "EU 'Homeland Security' Lacks Democratic Oversight, Says Watchdog," *EUobserver*, October 2, 2009, https://euobserver.com/justice/28759

16. Regulation (EU) No. 1291/2013 of the European Parliament and of the Council of 11 December 2013 Establishing Horizon 2020—the Framework Programme for Research and Innovation (2014–2020) and Repealing Decision No. 1982/2006/EC, *Official Journal of the European Union,* L 347, December 20, 2013, http://inea.ec.europa.eu/download/legal_framework/regulation_12912013_establishing_h2020.pdf
17. Ibid.
18. Ibid.
19. European Commission, "European Industry Leaders and EU Policymakers Meet to Plan for Security Research."
20. Piaguet and Gouardères, "Fact Sheets on the European Union—Defence Industry."
21. European Commission, "Horizon 2020 Work Programme 2014–2015," European Commission Decision C (2014)4995, July 22, 2014, http://ec.europa.eu/research/participants/portal/doc/call/h2020/common/1617621-part_19_general_annexes_v.2.0_en.pdf
22. Raymond B. Fosdick, "European Police Systems," *Journal of Criminal Law and Criminology* 6, no. 1 (1915): Article 3, http://scholarlycommons.law.northwestern.edu/cgi/viewcontent.cgi?article=1350&context=jclc
23. "The European Gendarmerie Force," European Gendarmerie Force, last accessed March 7, 2015, http://www.eurogendfor.org/organization/what-is-eurogendfor
24. "Community Acquis," *Wikipedia,* last published December 16, 2014, http://en.wikipedia.org/wiki/Community_acquis
25. Madeleine Vasquez et al., "Study on the Feasibility of Improved Co-operation between Bodies Carrying out European Coast Guard Functions," ICF International, Final Report, June 2014, http://ec.europa.eu/transport/modes/maritime/studies/doc/2014-06-icf-coastguard.pdf
26. Anne R. van Ewijk, "Diversity within Police Forces in Europe: A Case for the Comprehensive View," *Policing* 5 (2011).
27. "Taxation and Customs Union—The Customs Laboratories European Network (Overview)," European Commission, last published February 4, 2015, http://ec.europa.eu/taxation_customs/customs/customs_controls/customs_laboratories/group_ecl/index_en.htm
28. FRONTEX website, last accessed March 7, 2015, http://frontex.europa.eu
29. Decision No 1313/2013/EU of the European Parliament and of the Council of 17 December 2013 on a Union Civil Protection Mechanism, *Official Journal of the European Union,* L 347, December 20, 2013, http://eur-lex.europa.eu/LexUriServ/LexUriServ.do?uri=OJ:L:2013:347:0924:0947:EN:PDF
30. "EU's 'Triton' Operation Gets Underway in Mediterranean," Deutsche Welle, published November 1, 2014, http://www.dw.de/eus-triton-operation-gets-underway-in-mediterranean/a-18033932
31. EUROPOL website, last accessed March 7, 2015, https://www.europol.europa.eu

32. "More than 1,000 Arrested in Huge EU Crime Crackdown," *BBC News Europe,* published September 24, 2014, http://www.bbc.com/news/world-europe-29344979
33. Jane Wakefield, "Huge Raid to Shut Down 400-Plus Dark Net Sites," *BBC News Technology,* published November 7, 2014, http://www.bbc.com/news/technology-29950946
34. euLISA website, last accessed March 7, 2015, https://www.eulisa.europa.eu
35. "Emergency Response Coordination Centre (ERCC)," European Commission, Humanitarian Aid and Civil Protection, last published June 10, 2014, http://ec.europa.eu/echo/en/what/civil-protection/emergency-response-coordination-centre-ercc
36. Centre for Strategy and Evaluation Services, "Ex-Post Evaluation of the Preparatory Action on Security Research (PASR)," January 2011, http://ec.europa.eu/enterprise/policies/security/files/doc/interim_evaluation_of_fp7_security_ex_post_pasr_final_report_en.pdf
37. Commission of the European Communities, "Commission Communications. Implementation of Preparatory Action on the Enhancement of the European Industrial Potential in the Field of Security Research: Towards a Programme to Advance European Security through Research and Technology," COM(2004)72, February 3, 2004, http://eur-lex.europa.eu/LexUriServ/LexUriServ.do?uri=COM:2004:0072:FIN:EN:PDF; Commission Decision of 3 February 2004 on the Implementation of the Preparatory Action on the Enhancement of the European Industrial Potential in the Field of Security Research (2004/213/EC), *Official Journal of the European Union,* L 67, March 5, 2004, http://eur-lex.europa.eu/LexUriServ/LexUriServ.do?uri=OJ:L:2004:067:0018:0022:EN:PDF; Commission of the European Communities, "Communication from the Commission to the Council, the European Parliament, the European Economic and Social Committee and the Committee of the Regions: Security Research. The Next Steps," COM(2004) 590 final, September 7, 2004, http://eur-lex.europa.eu/LexUriServ/LexUriServ.do?uri=COM:2004:0590:FIN:EN:PDF
38. Finland, France, the United Kingdom, Sweden, Austria, Germany, and the Netherlands.
39. Regulation (EU) No. 1291/2013 of the European Parliament and of the Council of 11 December 2013 establishing Horizon 2020—the Framework Programme for Research and Innovation (2014–2020) and Repealing Decision No. 1982/2006/EC.
40. "European Network of Law Enforcement Technology Services," Police Department under the Ministry of the Interior of the Republic of Lithuania, Presidency of the Council of the European Union, last published July 16, 2013, http://www.policija.lt/index.php?id=21280
41. "Comparative Study: IPR-Related Provisions in the US and EU R&D Funding Regimes," unpublished study for the European Commission, 2014.
42. It should be noted that although the existence of "march-in rights" do make the U.S. legal regimes legally "tougher" on contractors, in practice the right is rarely invoked and has never been successful.
43. See *U.S. Code (*1952), *Title 35*, Sections 202—Disposition of Rights and 203—March-in Rights.

44. As defined in Article 20(1) of the Treaty on European Union (and further expanded in Title III, Articles 326–334 of the Treaty on the Functioning of the European Union, see Consolidated Versions of the Treaty on European Union and the Treaty on the Functioning of the European Union, *Official Journal of the European Union,* C 326, October 26, 2012, http://eur-lex.europa.eu/legal-content/EN/TXT/PDF/?uri=OJ:C:2012:326:FULL&from=EN
45. "Enhanced Cooperation," *Wikipedia,* last published February 12, 2015, http://en.wikipedia.org/wiki/Enhancedcooperation; "Enhanced Co-operation," *EUABC.com,* last accessed March 9, 2015, http://en.euabc.com/word/332; Carlo Maria Cantore, "We're One, but We're Not the Same: Enhanced Cooperation and the Tension between Unity and Asymmetry in the EU," *Perspectives on Federalism* 3, no. 3 (2011), http://on-federalism.eu/attachments/103download.pdf; Federico Fabbrini, *The Enhanced Cooperation Procedure: A Study in Multispeed Integration* (N.p.: Centro Studi sul Federalismo, 2012), http://www.csfederalismo.it/attachments/2433_CSF-RP_Fabbrini_ENHANCED%20COOPERATION_October2012.pdf
46. "Whereas activities will have an exclusive focus on civil applications, coordination with the activities of the European Defence Agency (EDA) may be considered with possible synergies being established with projects funded by the EDA programs. The complementarity of such synergies should be described comprehensively." European Commission, "European Commission Decision C (2014)4995 of 22 July 2014, Horizon 2020 Work Programme 2014–2015. 14. Secure Societies—Protecting Freedom and Security of Europe and Its Citizens. Revised," 13–14, http://ec.europa.eu/research/participants/data/ref/h2020/wp/2014_2015/main/h2020-wp1415-security_en.pdf
47. "Research and Technology Strategy," European Defence Agency, last accessed March 9, 2015, https://eda.europa.eu/Aboutus/Whatwedo/strategies/ResearchandTechnology
48. "Common European Security and Defence Policy," *Wikipedia,* last published January 2, 2015, http://en.wikipedia.org/wiki/Common_Security_and_Defence_Policy
49. For instance Statewatch (http://www.statewatch.org) and other organizations that are members of the European Civil Liberties Network (http://www.ecln.org)
50. "Research Ethics," European Commission, Research and Innovation, last accessed March 9, 2015, http://ec.europa.eu/research/swafs/index.cfm?pg=policy&lib=ethics
51. Regulation (EU) No. 1291/2013 of the European Parliament and of the Council of 11 December 2013 establishing Horizon 2020—the Framework Programme for Research and Innovation (2014–2020) and Repealing Decision No. 1982/2006/EC, Article 19(1) states: "All the research and innovation activities carried out under Horizon 2020 shall comply with ethical principles and relevant national, Union and international legislation, including the Charter of Fundamental Rights of the European Convention on Human Rights and its Supplementary Protocols."

52. Proposals need to comply with export rules, including technology transfers, as defined under Council Regulation (EC) No. 428/2009 of 5 May 2009, Setting Up a Community Regime for the Control of Exports, Transfer, Brokering and Transit of Dual-Use Items (Recast), *Official Journal of the European Union,* L 134, May 29, 2009, http://eur-lex.europa.eu/LexUriServ/LexUriServ.do?uri=OJ:L:2009:134:0001:0269:en:PDF
53. European Commission, "Guidance Notes on Project Reporting, FP7 Collaborative Projects, Networks of Excellence, Coordination and Support Actions, Research for the Benefit of Specific Groups (in particular SMEs). Version 2012, last accessed March 9, 2015, http://ec.europa.eu/research/participants/data/ref/fp7/89692/project-reporting_en.pdf, Annex 1
54. "Privacy by Design," *Wikipedia,* last published March 4, 2015, http://en.wikipedia.org/wiki/Privacy_by_Design

Part V

Conclusion and Outlook

13

The Scientific Status of New Security Studies: A Critical Search for Epistemic Identity of Homeland and Civil Security Research

RACHEL SUISSA

Introduction

This chapter puts the contributions to this volume into perspective, focusing on the status of homeland and civil security as a field of research. It argues that the epistemic identity of this field can be provided by the overarching academic discipline of *new security studies.* The conceptual radical shifts from *security* as defense and war studies, military, grand strategy, and geopolitics to comprehensive *new security studies,*[1] including *homeland and civil security,* require an epistemological inquiry (i.e., a study of the nature and grounds of the knowledge in the field) to help evaluate their scientific status. An evaluation of the scientific status of a research program and its theoretical core assumptions should be differentiated from the epistemology of the evolution of the discipline to which the research program relates. *New security studies* and their novelty are defined by the following four pillars,[2] each of which have been addressed by chapters in this book:

- *"New security concepts"*—hybrid threats (see Chapter 1), human security, civilizational security, comprehensive approach, resilience (see Chapter 4 and also Chapter 12), etc.;
- *"New security subjects"*—biopolitics of security, financial security, security as ethics (see Chapter 10), etc.;

- *"New security objects"*—environmental security, food security, cybersecurity, health and pandemic security, border security (see Chapter 2), etc.; and
- *"New security practices"*—migration and insecurity, security technologies (see Chapter 7), commercial security practices, including management of industrial accidents (see Chapter 3), professional security practices such as leadership (see Chapter 9), intelligence (see Chapters 5 and 6), risk management (see Chapter 8), etc., as well as academic teaching practices (see Chapter 11).

Thus, new security studies are being discussed mainly through their ontological entities, with valuable contributions to the state of the art, though sometimes at the risk of confusing security ontology (what the nature of security is) with questions of its epistemology (how we can gain knowledge about security).

Any inquiry into the scientific status of new security studies as an overarching discipline for scientific inquiry into homeland and civil security should acknowledge that security belongs to hermeneutics, which is the study of society rather than of nature or built environment. In our field, this is probably best exemplified by "securitization" theory, which assumes that security issues do not appear and dissolve because the threat environment changes, but because dominant discourses transform.[3] The securitization of the European Union's external border and the creation of FRONTEX, for example (see Chapters 2 and 12), have been interpreted in this sense.[4] Further, when pursuing the evolution of a discipline or field, its conceptualization in ontological terms is crucial. This involves both theoretical and practical reasons. However, if the aspect of practicality that dominates security as a branch of knowledge is the reason for inquiring on its scientific status, its ontological entities might be used mistakenly to identify science as equal to practice.

This is of special relevance when referring to *new security studies* as an overarching frame of reference for *homeland and civil security research,* which itself is multidisciplinary. Therefore, the chapter will first discuss distinctions made in the evolution of security studies elaborating ontological entities. A discussion of new security studies in the context of classical and anti-classical epistemology will follow in order to determine its place in different epistemic contours. The evaluation of the scientific status of this branch will be carried out through the examination of different "demarcation criteria," as discussed in the study and philosophy of science and to be addressed below.

Beyond One Discipline

Questioning the scientific status of a discipline cannot be detached from the *epistemic culture* its branch of knowledge belongs to. When raising this question in new security studies, the relevant concept of epistemic culture is associated with relativism, constructivism, and securitization.[5] Security is no longer limited to the traditional policies and military tools of the nation-state, and is no longer considered dependent primarily on geopolitics and military strength. Security has expanded into a broad, comprehensive field that encompasses social, economic, environmental, and ethical models of analysis and tools of action. This comprehensiveness exposes the security-cultural *epistemic relativism* to intersubjective levels when different entities differ in the *ontological* concept of *security* and the balance between models of tools and actions. It increases the relevance of different approaches to *constructivist analysis* of how security threats, policies, and strategies emerge within the *securitization* of everyday life across borders.

There is no reason to suspect that homeland and civil security as a specific subject matter and field of study should be exempt from those implications of the postmodern condition. Issues acquire a security status and become securitized as a result of an intersubjective process involving a particular discursive and political force that, once established, enables policy makers to immediately adopt whatever means they deem appropriate to curb the threat.[6] Informed by Popper's critical approach to historicism,[7] new security studies need to explore the linkage between the ontology of disciplines, the scientific methods they use, and their political implications.

Following Popper's demarcation criterion between science and pseudoscience, new security studies will only be a science if they are open to be disproven, or "falsified," by the facts. As the following example from homeland security shows, this is not a theoretical construct but of direct practical relevance. In order to protect against potential terrorist threats, it may be legitimate to take preventive measures when there is a valid threat, yet insufficient evidence of an impending terrorist attack. This must be distinguished from taking measures against an alleged terrorist threat that may be a theoretical possibility, but for which there is no valid evidence present. Therefore, decision-makers in homeland security must be able to distinguish between scientific and pseudoscientific claims.

Security acquires a particular discursive and political force as it becomes concept related to action—securitizing (or de-securitizing) something—rather to an objective (or subjective) condition.[8] This consensual assumption risks being accused of using epistemology as a normative political force to criticize security governance and the political management security problems. New security studies however entail much more than securitization studies. They as well reach beyond critical analysis of *"riskification,"* that is, how risk assessment is politically and socially managed rather than being threat-based, directly addressing causes of harm.[9] Rather, in broad terms, new security studies can be considered *holistic*: Within a particular research program, they are subjective, objective, and discursive; use different methods; address different levels and units of analysis; as well as use multi-method and multi-level analysis. New security studies are also *cross-disciplinary* in that perspectives from a variety of disciplines are actively involved. For example, security can be viewed through economics—as a perceived threat, like extreme energy-price volatility, global imbalances and currency volatility, infrastructure fragility, and regulatory failures. It can also be viewed in relation to national security and national power. In addition, new security studies have *multidisciplinary* characteristics because they involve people from different disciplines working on the same security topics to address them comprehensively, each contributing their specific disciplinary knowledge to the common effort.[10]

This is particularly relevant in homeland security practice and civil security research, which both are based on an all-hazards approach. Using the study of *vulnerability* as an example, the following Table 13.1 illustrates a holistic view on this topic by summarizing different relevant disciplinary perspectives and the practical recommendations that follow for vulnerability-reducing measures:

Table 13.1. Vulnerability in Multidisciplinary Perspective.

Discipline	View(s) of Vulnerability	Recommendation(s)
Geography	Vulnerability is determined by the use of hazard-prone areas	Land-use planning that takes into account hazards to reduce risk
Meteorology	Vulnerability is due to a lack of advanced warning of severe weather	Acquisition, creation, and effective use of warning systems
Engineering	Vulnerability occurs when structures and infrastructure cannot withstand the forces of hazards	Design and construction of buildings and infrastructure that promotes disaster resistance
Anthropology	Vulnerability emanates from constraining values, attitudes, and practices	Alter attitudes to discourage risk-taking practices and susceptibility

Discipline	View(s) of Vulnerability	Recommendation(s)
Economics	Vulnerability is related to poverty and results in an inability to prevent, prepare for, or recover from a disaster	Improve the distribution of wealth and purchase insurance to minimize losses and promote resilience
Sociology	Vulnerability is a product of inaccurate assumptions about disaster behavior and is related to race, gender, age, disability, etc.	Understand behavioral patterns in disasters and pay attention to needs of special populations
Psychology	Vulnerability is a function of overlooking or minimizing risk and not being able to cope emotionally with stress and/or loss	Help people to recognize risk and provide crisis counseling to enable resilience
Epidemiology	Vulnerability is susceptibility to disease or injury and is related to malnutrition and other health factors	Improve provision of public health/emergency medical care before, during, and after disasters
Environmental Science	Vulnerability is proneness to environmental degradation, which may change weather patterns and produce long-term disasters	Conserve natural resources, protect green space areas, and ensure that debris management is performed in an environmentally conscious manner
Political Science	Vulnerability is produced by the political structure and incorrect decision making	Alter structure of political system and educate politicians and legislators about disasters
Public Administration	Vulnerability results from misguided laws, the failure to implement policies effectively, and an inability to enforce regulations	Strengthen response and recovery capabilities through preparedness measures, improved policy implementation and increased code enforcement
Law	Vulnerability results from negligence, which is a failure to act as reason or legal statutes dictate	Understand the law, alter statutes, and ensure compliance to widely accepted ethical practices in emergency management

Discipline	View(s) of Vulnerability	Recommendation(s)
Journalism	Vulnerability is a result of insufficient public awareness about hazards and how to respond to disasters	Dispel myths about disasters, foster increased media capabilities, and educate the public about hazards
Emergency Management	Vulnerability is the lack of capacity to perform important functions before and after disaster strikes (e.g., evacuation, search and rescue, public information, etc.)	Foster public awareness about disasters and build capacities through hazard and vulnerability analyses, resource acquisition, planning, training, and exercises
Homeland Security	Vulnerability is due to cultural misunderstandings, permeable borders and fragile infrastructure, and weak disaster management institutions	Correct domestic and foreign policy mistakes, enhance counterterrorism measures, protect borders and infrastructure, and improve WMD capabilities

Source: David A. McEntire, "The Importance of Multi- and Interdisciplinary Research on Disasters and for Emergency Management," FEMA Emergency Management Institute, last accessed May 3, 2015, 15, https://training.fema.gov/HiEdu/docs//EMT/Chapter%20-%20Introduction.doc

In an ideal stage of evolution, new security studies would be *interdisciplinary*, integrating knowledge and methods from different disciplines by synthesizing approaches: for instance, integrating political science in homeland security decision-making thought, processes, and institutions (see Chapter 11); or integrating computer sciences and information management in cybersecurity strategy and cyber preparedness. That way, new security studies could become an overarching academic discipline for homeland and civil security research.

New Security Studies between Classical and Anti-Classical Epistemology

Classical epistemology that is concerned with the pursuit of truth with rationality or epistemic justification cannot serve the security researcher. This approach relates to an *individual* engaging in cognitive activity to arrive at a true belief and to avoid false belief. However, security studies cannot be described as a branch of an epistemology in its *individualistic* guise restricted to truth or justification.

In homeland and civil security research that rests on the vision of providing scientific knowledge to improve the creation of security as a public good and its delivery to the whole community of citizens, social epistemology—the study of the social dimensions of knowledge or information—is an important framework for security knowledge to be evaluated. New security studies thus belong to the branch of *social epistemology* that studies the evolution of security knowledge within social contexts and practices, such as securitization.

Social epistemology does not have a long history of systematic use. However, Plato, in his dialogue *Charmides,* can be seen as one who made a brief foray into the social dimensions of knowledge or rational belief. Plato posed the question of how a layperson can determine whether someone who purports to be an expert in an area really is one. The seventeenth and eighteenth century British philosophers John Locke, David Hume, and Thomas Reid used the term "testimony," asking, "When should cognitive agents rely on the opinions and reports of others?" and "What must a listener know about a speaker to be entitled to trust his assertions?" Locke's intellectual self-reliance doubted giving authority to the opinions of others. Hume took reliance on factual statements of others for granted only when there are adequate reasons for thinking that these sources are reliable. Hume's empiricism was grounded on personal observations that establish the veracity of human testimony. Reid, by contrast, related to our natural attitude of trusting others as reasonable even if little is known about their reliability.[11]

These epistemological positions are all practically relevant when describing the communicative profile of security knowledge. Threat detection and threat perception involve intersubjective relationship between experts/authorities/testimony and laypersons/listeners. Threat detectors differ in threat perceptions. Their audience might either doubt their assessment due to their intellectual self-reliance, might reject/accept the empirical observations that support their assessment, or even trust without proof.

Security knowledge also seems to fit into a different tradition, which focuses on aspects of knowledge that are *social* in a sociological or political sense. Karl Marx's theory of ideology could well be considered a type of social epistemology in which "ideology" is a set of beliefs, a worldview, or a form of consciousness that in some respects is false or delusive. These beliefs and their delusiveness derive from the social situation and interests of the believers.[12] In contrast, many approaches have focused on security as constituting something that needs to be secured: the nation, the state, the homeland, the infrastructure, etc.—almost to the extent of necessitating the conclusion that none of those would even exist if they did not need to be secured.

The *critical theory* of the Frankfurt School proclaimed, "[c]ritical theory aims at emancipation and enlightenment by making agents aware of hidden coercion in their environment, enabling them to determine where their true interests lie."[13] Jürgen Habermas, one of their exponents, introduced the concept of "ideal speech situation," a hypothetical situation of absolutely uncoerced and unlimited discussion between completely free and equal human agents, and described this speech situation as a transcendental criterion of truth. Though this notion of perfect cognitive symmetry is a valid epistemic standard, it cannot be applied to security knowledge, or to the assessment of its scientific quality. While critical thinking is important in all fields of security studies, based on its own set of research methods,[14] the generic properties of security knowledge cannot be reduced to the universal standard posited by critical theory, judging all knowledge on the basis of symmetry of power between speaker and listener. The reason for this is that security is about leadership (see Chapter 9) and involves differences in professional power and expertise reflected in the social epistemology of knowledge. Again, the argument is not that there is no place for critical thinking when it comes to security, but that the absence of the "ideal speech situation" does not preclude security knowledge being able to be formulated as scientific knowledge.

New security studies therefore have an anti-classical, social epistemology that rejects the classical concerns of epistemology with truth, justification, and rationality and seeks to understand a selected community's norms of rationality. It also rejects the notion that this discipline of knowledge has universal or "objective" norms of rationality, or criteria of truth, that it could appropriately invoke. Security studies are neither context-free nor do they establish norms of rationality without cultural roots and limitations. At the same time, the epistemic effort of new security studies cannot be isolated from disciplines in which rationality is a natural objective norm. For instance, analyzing security threats that stem from environmental disciplines might involve molecular biology, genetics, chemistry, etc., that do evoke rationality as a legitimate objective norm. Another example is the theorizing process in cybersecurity that integrates different technology disciplines to various problematic fields of security. Homeland and civil security research cannot be detached from its transdisciplinary profile, and the question of the scientific status of a research program in security studies, in this case, is the scientific status of a transdisciplinary set of knowledge.

A crucial challenge for new security studies as a transdisciplinary research process relates to the complexity of the problems it seeks to understand: the diversity of life; scientific perceptions of problems it takes into account; the link to abstract and case-specific knowledge; and the development of knowledge and practices that promote what is perceived to be the common good.[15] In this process, engaging research and collaboration between disciplines are the means of meeting all these requirements. Thus, if the evaluation of scientific status were to be made according to the demarcation between *science* and *pseudo-science,* through *informal logic* and *rational argumentation*, this challenge could possibly be overcome. Due to the practical implications of the transdisciplinary study of security, the demarcation criterion is important for decision support and we need to distinguish its scientific knowledge from its look-alikes. Because of the high relevance of security in today's societies, the need for demarcation is not only important for the validity of the social epistemology of security research, but also of high practical relevance.

However, although the demarcation criterion could be applied to evaluate the scientific status of a research program in security studies on the grounds of practical benefits one may gain, such an approach is opposed to Popper's demarcation based on empirical falsification. According to Popper, a theory—or in a broader sense the validity of knowledge—in the empirical sciences can never generally be proven, but it can only be falsified on a case-by-case basis, meaning that it can and should be scrutinized using specific experiments. Falsifying a theory that has reached its transformation to practical applications could add to the challenge of evaluating a transdisciplinary research program for homeland and civil security and its underlying assumptions. Therefore, a careful analysis of the status of new security studies through the demarcation criterion is necessary.

New security studies cannot be reduced to a single branch of knowledge, due to their comprehensive and transdisciplinary profile, and its proximity to practice and co-evolution with real-world security policies and programs raises the question of its scientific status. This needs clarification. The English word "science" is primarily used for the natural sciences and other fields of research that are considered similar to them. Hence, political economy and sociology are considered sciences, whereas studies of literature and history are not. The corresponding German word "Wissenschaft" covers all academic specializations, including the humanities, and has the advantage of more adequately delimiting the type of systematic knowledge that is in dispute in the conflict between science and pseudo-science. New security studies, like the natural sciences, social sciences, and the humanities, are part of the same human endeavor, namely, systematic and critical investigation aimed at acquiring the best possible understanding of the workings of nature, people, and human society.

The disciplines that form this community of knowledge are increasingly interdependent. Integrative disciplines such as evolutionary biology, biochemistry, ecology, the neurosciences, and game theory play a role in conceptualizing new security studies today and thus contribute to linking previously unconnected disciplines.[16] These interconnections have brought the sciences and the humanities closer to each other. Evolutionary biology theory, for instance, has been considered as evolutionary thinking in global challenges in the areas of food security, emerging diseases, and resilience. Biochemistry contributes to this challenge, to the industrial manufacture of strategic weapons and even to relevant ethical dilemmas of the Biological and Toxin Weapons Convention (BWC) and the Chemical Weapons Convention (CWC). Improving the latter cannot be done without biologists, chemists, and researchers from a variety of other disciplines whose work helps assess dual-use implications, including researchers from the humanities.

The question remains of what demarcation criteria should be applied to assess the scientific status of new security studies. Should they be evaluated through a research program, an epistemic field, or cognitive discipline (i.e., a group of people with common knowledge aims and practices), a theory, a practice, or a scientific inquiry? The answer is that new security studies cannot be reduced to any one of these levels, as argued in the following.

The Scientific Status of New Security Studies under Alternative Demarcation Criteria

Although focused on international and strategic security rather than homeland and civil security, the new edition of Barry Buzan's book *People, States, and Fear: An Agenda for International Security Studies in the Post-Cold War Era* (2009) provides a model for our discipline as well, since it offers constructivist and ethical reflections that no longer reduce security as such to positivist and rationalist epistemologies.[17] New security studies as an overarching academic discipline for the analysis of homeland and civil security in fact follow a constructivist and interpretive epistemology, with new quality criteria for its knowledge, such as being plausible; meeting social needs; underpinning values; helping to build and maintain communities; etc.—in addition to the traditional, rational(ist) criteria of validity, reliability, and objectivity. Based on this, the following four demarcation criteria are proposed to determine the scientific status of new security studies: *falsificationism*; *puzzle solving*; *scientific progress*; and *multi-criteria approaches*.

Falsificationism was introduced by Popper, who rejected, as mentioned above, verifiability as a criterion for a theory or hypothesis to be scientific, rather than pseudo-scientific or metaphysical. To be ranked as scientific, a theory should be falsifiable and must be capable of conflicting with possible or conceivable observations. However, a theory of security is almost impossible to refute or falsify. New security studies have become a diverse branch of knowledge in which a weak and limited theory can never really be thrown away. Once *verified* under limited conditions, it will typically become part of the established set of knowledge and assumptions and be followed by researchers. It may even resist logical and practical falsification, as arguably security policy may as well.

This also applies to Thomas Kuhn's criterion of *puzzle solving*.[18] Kuhn divided the process of doing science into "normal science" and "extraordinary science." In normal science, the scientist's activity consists of solving puzzles rather than testing fundamental theories. In puzzle solving, current theory is accepted, and the puzzle is defined in its terms. Kuhn's own demarcation criterion is the capability of puzzle solving, which he sees as an essential characteristic of normal science. However, since Kuhn's view develops into paradigm shifts in which a jettisoned paradigm is atrophied when no longer useful for solving puzzles, it becomes similar to Popper's falsificationism. The only difference is in the volume of the research scope. Popper refers to the scope of a theory, while Kuhn refers to the scope of a paradigm. Since both can be invalid as different demarcation criteria, they cannot be used to demark new security studies as a scientific or pseudo-scientific branch of knowledge. A good example is the development of the different generations of dominant paradigms in security studies, such as the post-behaviorist and constructivist approaches that followed the behaviorist one. Behaviorism is still relevant to new security studies, not as a dominant paradigm but as one adapted alongside other paradigms. In homeland security and emergency management for example, assumptions on response to perceived threat (such as evacuation behavior) as well as models for societal resilience still rest on behaviorist reasoning.

The criteria based on *scientific progress* follow Imre Lakatos, who rejects Popper's demarcation criterion that focuses on the theory level and expands it to become "sophisticated" (methodological) falsificationism.[19] Instead of applying it to an isolated hypothesis or theory, he refers to this criterion as a whole research program that is characterized by a series of theories successively replacing each other. In his view, new security studies can be considered as a scientific branch if its progress is in new theories that make surprising predictions that are confirmed. In contrast, they would be considered a degenerating research program if characterized by theories being fabricated only in order to accommodate known facts:

> Progress in science is only possible if a research program satisfies the minimum requirement that each new theory that is developed in the program has a larger

> empirical content than its predecessor. If a research program does not satisfy this requirement, then it is pseudo-scientific.[20]

According to Lakatos' demarcation criterion, new security studies in general, and homeland and civil security research in particular, are scientific epistemic fields. They are far from being described as a degenerating research program. Since they possess the requirement of being adequately integrated into other sciences and are incorporated into the existing network of established sciences, they are fully in line with the criteria of a progressive research program. Lakatos's demarcation criterion can be used to identify new security studies, including homeland and civil security research, as a science when combined with the following criteria to rule out pseudo-science or pseudo-scientific practice. According to Mario Bunge, pseudo-science may be "diagnosed" using the following criteria:[21]

- *Belief in authority*: It is contended that some person or persons have a special ability to determine what is true or false. Others have to accept their judgments.
- *Unrepeatable experiments*: Reliance is put on experiments [or quasi-experiments and case studies] that cannot be repeated by others with the same outcome.
- *Handpicked examples*: Handpicked examples are used although they are not representative of the general category that the investigation refers to.
- *Unwillingness to test*: A theory is not tested although it is possible to test it.
- *Disregard of refuting information*: Observations or experiments that conflict with a theory are neglected.
- *Built-in subterfuge*: The testing of a theory is so arranged that the theory can only be confirmed, never disconfirmed, by the outcome.
- *Explanations are abandoned without replacement*: Tenable explanations are given up without being replaced, so that the new theory leaves much more unexplained than the previous one.[22]

Our research enterprise mainly is co-evolving with the mission space of the real-world homeland security enterprise, to which it seeks to offer problem solving as well as critical thinking related to policies and strategies aiming at averting harm from societies. It thus can be identified as: a scientific field that progresses through new assumptions and theories that can be confirmed by current reality, but that might not be repeated by others; might be resistant to more thorough testing as long as they seem to work in reality, with practical value; and might neglect conflicting observations and stick to theories and assumptions that can only be confirmed, but not refuted (or falsified).

While the grant-driven and end-user needs-based evolution of security research may be seen to add to these risks of making new security studies a pseudo-science, the strong focus of security research on tangible results and consortia that are multidisciplinary by design and involve industrial research and development (see Chapter 12) certainly adds to the scientific status of the field. While on a good way, we as a community of scholars need to maintain the momentum in our effort to avail homeland and civil security research a valid epistemic identity within a field of new security studies that is consolidating its scientific status. At the same time, we need to bear in mind that while there is a set of features that are characteristic of science, every part of science will have some of these features, but not any part of science can be expected to have them all.[23]

Conclusion

The contributions to this volume have demonstrated how cross-disciplinary perspectives—including practical accounts of single cases and threat environments, as well as comparative studies of homeland security policies—can collude to foster the common goal of advancing a common epistemic field and share practices of scientific inquiry (see Part I of this book, Chapters 1–4). This chapter has discussed how the scientific status of the evolving field can be evaluated according to several demarcation criteria as they derive from major results of the study of science. The recent emphasis of transversal aspects of the field and of critical thinking (see Part III of this book, Chapters 9–10), including a critical perspective on the inquiries' own limitations and risks (e.g., ethical issues as addressed in Chapter 10), roots homeland and civil security research still stronger within the overarching discipline of *new security studies*. To further forge its epistemic identity supported by forward-looking programming (as addressed in Part IV of this book, Chapters 11–12), and maintain its scientific status and its utility (i.e., to remain practically relevant), homeland and civil security research will need to maintain multi- and cross-disciplinary approaches (as exemplified in Part II of this book, Chapters 5–8) to vulnerability assessment, to resilience enhancement, and to many other objectives, both scientific and practical. It will also need to remain open to reviewing the epistemology of the homeland and civil security field of research as well as that of the evolving real-world mission space, such as the main assumptions about the threats that constitute the homeland security enterprise and the civil security model.

Notes

This chapter has been supported by the Haifa Center for German and European Studies, University of Haifa.

1. J. Peter Burgess, ed., *The Routledge Handbook of New Security Studies* (Milton Park: Routledge, 2010).
2. Ibid., own examples added.
3. Constructivism and the Copenhagen School's theory of securitization are particularly relevant here. A seminal source is Thierry Balzacq et al., "Forum: What Kind of Theory—If Any—Is Securitization?" *International Relations*, first published on October 21, 2014 as doi:10.1177/0047117814526606, http://ire.sagepub.com/content/early/2014/10/27/0047117814526606.full.pdf+html; see also Thierry Balzacq, ed., *Securitization Theory: How Security Problems Emerge and Dissolve* (London and New York: Routledge, 2011).
4. Andrew W. Neal, "Securitizing and Risk at the EU Border: The Origins of FRONTEX," *Journal of Common Market Studies* 47 (2009).
5. See Mike Bourne, *Understanding Security* (Houndmills, Basingstoke: Palgrave Macmillan, 2014), chapter 3: "Critical Approaches and New Frameworks," 50–70.
6. Barry Buzan, *People, States & Fear: The National Security Problem in International Relations* (Chapel Hill: University of North Carolina Press, 1983); Barry Buzan and Lene Hansen, *The Evolution of International Security* (Cambridge: Cambridge University Press, 2009); Balzacq, *Securitization Theory*; Balzacq et al., "Forum: What Kind of Theory—if Any—Is Securitization?"
7. Karl Popper, *The Open Society and Its Enemies* (London: Routledge, 1945).
8. Balzacq, *Securitization Theory.*
9. Olaf Corry, "Securitisation and 'Riskification': Second Order Security and the Politics of Climate Change," *Millennium: Journal of International Studies* 40 (2011).
10. David A. McEntire: "The Importance of Multi- and Interdisciplinary Research on Disasters and for Emergency Management," in *Disciplines, Disasters and Emergency Management: The Convergence and Divergence of Concepts, Issues and Trends from the Research Literature,* ed. David A. McEntire (Springfield, IL: Charles C. Thomas, 2007), 6 [also available in an e-book version: Emmitsburg, MD: FEMA Emergency Management Institute, https://training.fema.gov/hiedu/aemrc/booksdownload/ddemtextbook]

11. Alvin Goldman, "Social Epistemology," in *Stanford Encyclopedia of Philosophy*, ed. Edward N. Zalta, last published August 18, 2006, chapter 1, http://plato.stanford.edu/entries/epistemology-social
12. Ibid.
13. Ibid.
14. See Mark B. Salter and Can E. Mutlu, eds., *Research Methods in Critical Security Studies: An Introduction* (New York and London: Routledge, 2013); Laura J. Shepherd, ed., *Critical Approaches to Security: An Introduction to Theories and Methods* (New York and London: Routledge, 2013).
15. Gabriela Wuelser, Christian Pohl, and Gabriela Hirsch Hadorn, "Structuring Complexity for Tailoring Research Contributions to Sustainable Development: A Framework," *Sustainability Science* 7 (2011).
16. See CEUSS | Center for European Security Studies, Sigmund Freud University Vienna, "Roadmap Proposal for the Development of a Trans-disciplinary Security Research Paradigm," deliverable 8.3, EU security research project FOCUS (Foresight Security Scenarios—Mapping Research to a Comprehensive Approach to Exogenous EU Roles), March 2013, http://www.focusproject.eu/documents/14976/63ec147d-c599-4d61-b2c7-be3d4590d3ea
17. Barry Buzan, *People, States & Fear: The National Security Problem in International Relations* (Boulder, CO: Rienner, 2009).
18. See Thomas Kuhn, *The Structure of Scientific Revolutions* (Chicago, IL: University of Chicago Press, 1962).
19. See Imre Lakatos and Alan Musgrave, eds., *Criticism and the Growth of Knowledge* (Cambridge: Cambridge University Press, 1965).
20. Sven Ove Hansson, "Science and Pseudo-Science," in *Stanford Encyclopedia of Philosophy*, ed. Edward N. Zalta, last published February 10, 2014, chapter 4.4, http://plato.stanford.edu/entries/pseudo-science, http://plato.stanford.edu/entries/pseudo-science
21. Mario Bunge, "Diagnosing Pseudoscience," in *Philosophy in Crisis. The Need for Reconstruction*, by Mario Bunge (Amherst, NY: Prometheus Books, 2001).
22. Hansson, "Science and Pseudo-Science," chapter 4.6.
23. Cf. ibid.

Glossary

All-hazards approach: The all-hazards approach is concerned with protecting infrastructure and society from the whole range of hazards (anthropogenic and natural), though not necessarily each and every single hazard.

Anthropogenic: "Man-made" (e.g., anthropogenic vs. natural disasters).

Big data: A broad term for data sets so large or complex that traditional data processing applications are inadequate. Challenges include predictive analysis, capture, curation, search, sharing, storage, transfer, visualization, and information privacy.

Black swan: An improbable event that has three main characteristics: unpredictability (due to lack of imagination of the "impossible" and of scientifically computable probability); crucial and severe impact; and people's endeavor for explanation after its occurrence.

Border control: The regulation of access of people and goods to a country via land, maritime, and air borders and ports of entry.

Border management: A broad concept typically including (1) border control; (2) coping with immigrants, asylum claimants, and refugees; (3) enhancing security and coping with cross-border threats; and (4) coordination between and among domestic and foreign agencies.

Chilling effect: Reluctance of people to take part in certain activities caused by fear that such participation will bring them under official, yet unreasonable, suspicion of criminal, terrorist, etc., activity.

Civil protection: The protection of people, the environment, and property in the event of anthropogenic, technological, and natural risks or emergencies. While in some countries considered a concept overcome by developments after the Cold War and integrated into the homeland security enterprise (which includes emergency management), the concept is still current and informs policies and programs in other countries, including in the EU.

Civil security: A broad concept to summarize security challenges to societies, and how to respond to them, after the end of the Cold War and by means other than defense only (but including dual-use capabilities), based on the idea of the creation of security as a public good. In some countries, civil security is the name of departments or agencies in charge of homeland security, including border security and emergency management. In other cases, such as in the EU, civil security summarizes all policies, strategies, and programs to respond to all-hazards threats to societies, except the defense dimension but including use of military assets. Civil security thus includes civilian security but reaches beyond.

Comprehensive approach: Aims at overarching solutions to security problems, with broad effects based on complementarity of actors, while considering all available options and capabilities, as well as the normative end-state of the security of society as a whole, based on a whole-of-community approach. The National Strategy for Homeland Security of the U.S. (2007) called for a comprehensive approach that included fostering a culture of preparedness, integrated risk assessment, and improved use of science and technology, among other aspects. The term also refers to covering the whole crisis management cycle in homeland and civil security: prevention, protection, mitigation, response, and recovery. The 2010 Quadrennial Homeland Security Review defined a comprehensive approach to homeland security by the following three elements: security (protecting the United States and its people, vital interests, and way of life); resilience (individual, community, and system robustness, adaptability, and capacity for rapid recovery); and facilitation and enforcement of lawful trade, travel, and immigration.

Criminal intelligence: Information collected by local, state, and federal government entities in order to stop crime and affect arrests. This information is used by law enforcement agencies for all levels of criminal activity.

Critical infrastructure: Systems and assets, physical or virtual, that support vital societal functions and the disruption or destruction of which would have a significant impact on a country as a result of the failure to maintain those functions.

Critical infrastructure protection: All activities aimed at ensuring the functionality, continuity and integrity of critical infrastructures in order to deter, mitigate, and neutralize a threat, risk, or vulnerability.

Cybersecurity: Cybersecurity mainly refers to the protection of cyber infrastructure. Cyber infrastructure means both hardware in the strict sense and the global super network of more than a billion computers as well as mobile devices and radio networks, mutually linked together and forming the Internet or the cyberspace.

Danger: A condition/situation out of which damage, harm, or injury could result.

Defense Support of Civil Authorities (DSCA): The process by which U.S. military assets and personnel can be used to assist in missions normally carried out by civil authorities.

Disaster: A serious disruption of the functioning of a community or a society causing widespread human, material, economic, or environmental losses that exceed the ability of the affected community or society to cope using its own resources. A disaster results from a combination of hazards, vulnerability, and insufficient risk management.

Domestic intelligence: Information collected by a government entity concerning the potentially damaging plans, intentions, actions, etc., of a group or individual within its own country. This type of information is normally used for counterterrorism purposes.

Domestic terrorism: Any act of violence that is dangerous to human life or potentially destructive of critical infrastructure or key resources committed by a group or individual based and operating entirely within a country, without direction or inspiration from abroad.

Dual-use items: Those items, including software and technology, which can be used for both civil and military purposes, including all goods that can assist in the manufacture of nuclear weapons or other nuclear explosive devices.

Electronic intelligence (ELINT): Electronic emissions picked up from modern weapons and tracking systems.

Emergency management: As defined in *International Standard ISO 22320, Societal Security—Emergency Management—Requirements for Incident Response* (2011), emergency management is based on a comprehensive risk management approach to prevention, preparedness, response, and recovery before, during, and after potentially destabilizing or disruptive events.

Emergency Management Assistance Compact (EMAC): Signed by all 50 states in the U.S., the District of Columbia, Puerto Rico, Guam, and the U.S. Virgin Islands in 1996, it provides specific clarity and details on providing assistance to other states when an emergency has been declared by a governor but has not reached the presidential declaration stage.

European Agency for the Management of Operational Cooperation at the External Borders of the Member States of the European Union (FRONTEX): An EU agency with the mission is to coordinate, reinforce, and streamline operational cooperation between EU Member States, Schengen Associated Countries, and other partners in order to enhance the external border security of the Union.

European Border Surveillance System (EUROSUR): An information-exchange system designed to improve management of the EU external borders. It enables near real-time sharing of border-related data between members of the network, consisting of Schengen countries and FRONTEX.

European Security Model: An integrated EU security architecture containing common tools for collaboration between law enforcement authorities and judicial bodies within the Member States, defined in the EU Internal Security Strategy (2010). It also includes addressing the causes of insecurity and not just its effects, with priorities on prevention across sectors (political, economic, social, etc.).

Foreign intelligence: Information relating to the capabilities, intentions, and/or activities of foreign governments or elements thereof, foreign organizations, foreign persons, or international terrorist activities. This information can be used in forming the collecting government's foreign policy, counterterrorism strategy, economic policy, etc.

Foreign Intelligence Surveillance Act (FISA): The FISA Act of 1978, several times amended after September 11, 2001, requires the U.S. Government, before it commences certain kinds of intelligence gathering operations within the United States, to obtain a judicial warrant similar to that required in criminal investigations.

Fusion center: A fusion center is a location (usually physical, but can be virtual as well) where officials from local, state, and federal governments, often along with members from the private sector, work alongside each other to ensure any pertinent intelligence is being shared across the spectrum of those with responsibilities for securing the homeland.

GALILEO: The Galileo program is Europe's initiative for a state-of-the-art global satellite navigation system, providing a highly accurate global positioning service under civilian control. Galileo will provide Europe with independence in satellite navigation, but will also be interoperable with GPS and GLONASS, the two other global satellite navigation systems.

Geographic information systems (GIS): A term that describes the hardware, software, geographic data, and people that enable geospatial information processes for making decisions.

Geographic information technologies (GIT): A complex term for a large array of tools, grouping geographic information systems (GIS), together with global positioning systems, remote sensing techniques, and other spatially related tools for decision-making.

Geospatial intelligence (GEOINT): Defined by the intersection of the intelligence tradecraft, geographic science, and geographic information technology, GEOINT means use of imagery and information for descriptive and predictive analysis of activities referenced to the place on Earth where they take place.

Hazard: Generally, a situation that poses a level of threat to life, health, property, or environment. Specifically, a set of theoretical maximum impacts that are expected in a given place in a specified time interval with a certain probability. A hazardous situation that has happened is called an incident. A hazard expresses the potential for disaster in a given area. Hazard and possibility interact together to create risk.

Holistic approach: A holistic approach builds on multifunction, such as linking security and environmental protection, security and (resource) efficiency, etc., in one single package of measures and solutions.

Home affairs: A term in particular used in countries with a British Commonwealth tradition, but also, for example, in the EU, to denominate department or agency responsibilities related to internal security, including homeland and civil security, disaster management, justice, and border and immigration.

Homegrown violent extremist: A person of any citizenship who has lived or operated primarily in a given country and advocates, is engaged in, or is preparing to engage in ideologically-motivated terrorist activities supporting the objectives of a terrorist organization, but acting independently of direction by such an organization.

Homeland defense: The protection, under the responsibility of the Department of Defense, of U.S. sovereignty, territory, domestic population, and critical infrastructure against external threats and aggression, or other threats as directed by the President of the United States of America.

Homeland security: Originally defined as a concerted national effort to prevent terrorist attacks, reduce a nation's vulnerability to terrorism, and minimize the damage and recover from attacks that do occur, homeland security today means a whole-of-community, all-of-government, and all-hazards approach to manage risks to societies and prevent, protect against, mitigate, respond to, and recover from anthropogenic and natural disasters by using law-enforcement as opposed to military means, but including the use of supporting military capabilities.

Homeland security cycle: The adaptation of the common crisis management cycle to the homeland security mission space, as per Presidential Decision Directive 8 (PPD-8) on National Preparedness, with the phases of prevention, protection, mitigation, response, and recovery.

Human intelligence (HUMINT): The collection of information from human sources. The collection may be done openly or via clandestine or covert means (espionage). Within the United States, HUMINT collection is the FBI's responsibility. Beyond U.S. borders, HUMINT is generally collected by the CIA, but also by other U.S. components abroad.

Imagery intelligence (IMINT): Was practiced to a greater extent in World Wars I and II when both sides took photographs from airplanes. Today, in the U.S., the National Reconnaissance Office designs, builds, and operates imagery satellites, while the National Geospatial-Intelligence Agency is largely responsible for processing and using the imagery.

Intelligence cycle: The intelligence cycle describes intelligence collection as a process that is most commonly broken down into the steps of planning and direction; collection; processing; analysis and production; and dissemination.

Integrated approach: Focuses on cross-sector solutions based on platforms, such as providing security of both persons and goods in public transport by unified models, strategies, and technologies, or such as comprising different sectors of the homeland security cycle (prevention, protection, mitigation, response, and recovery).

Joint Terrorism Task Forces (JJTFs): Coordinated under the auspices of the Federal Bureau of Investigation, JTTFs consist of members from local, state, and federal law enforcement/security/intelligence agencies. The members of the JTTF work together for counterterrorism purposes.

Measurement and Signatures Intelligence (MASINT): A collection discipline that concerns weapons capabilities and industrial activities. MASINT includes the advanced processing and use of data gathered from overhead and airborne IMINT and SIGINT collection systems.

Military intelligence: Information collected by militaries and governments on the capabilities, plans, intentions, actions, etc., of another government's military. This information is used to form a country's military strategy.

Mission creep: The, often subtle, expansion of a project or mission beyond its original goals.

Moral hazard: Moral hazard may occur because the more secure people feel when engaging in certain activities, the more recklessly they behave, thereby increasing the risk to themselves of harm.

National Incident Management System (NIMS): A standardized approach to incident management developed by the U.S. Department of Homeland Security. It includes: a comprehensive, nationwide systematic approach to incident management; a core set of doctrine, concepts, principles, terminology, and organizational processes for all hazards; essential principles for a common operating picture and interoperability of communications and information management; standardized resource management procedures for coordination among different jurisdictions and organizations; and scalable and applicable for all incidents.

National Preparedness Goal (NPG): Established by Presidential Policy Directive 8 (PPD-8): National Preparedness (2011), it describes the U.S. nation-wide approach to achieving a secure and resilient nation, based on core capabilities, to preparing for the threats and hazards that pose the greatest risk to the security of the country. It defines national preparedness as a shared whole-of-community responsibility, including individuals, communities, the private and nonprofit sectors, faith-based organizations, as well as federal, state, and local governments.

National Response Framework (NRF): A guide to how the U.S. as a nation responds to all types of disasters and emergencies. Built on scalable, flexible, and adaptable concepts identified in the National Incident Management System (NIMS) to align key roles and responsibilities across the nation.

National security: A concept that a government, along with its parliament(s), uses to strategically prepare for and develop capabilities to protect the country and its citizens against all kinds of "national" crises through a variety of power projections, such as political power, diplomacy, economic power, military might, etc.

National Terrorism Advisory System (NTAS): Replacing the color-coded Homeland Security Advisory System (HSAS) in April 2011, the NTAS system effectively communicates information about terrorist threats to the U.S. by providing timely, detailed information to the public, government agencies, first responders, airports and other transportation hubs, and the private sector.

Need to know: A principle followed by the intelligence community, where agencies/departments share information via fusion centers or other means, but are often careful not to excessively disclose operational details, in order to avoid unnecessary risk for a specific operation or investigation.

New social media: Computer-mediated tools that allow people to create, share, or exchange information, ideas, and pictures/videos in virtual communities and networks.

New social network technologies: Those technologies include new social media such as Facebook, Xing, and Twitter, sharing facilities such as YouTube and Flickr, blogs, and dedicated websites; Web 2.0 technologies for collaborative applications that do not focus on social networking but on distributed volunteering etc., such as web applications for crowd sourcing (like Ushahidi.com); and the ubiquity of smartphones and their conceivable alternative uses for purposes of civil security, for example, based on disaster preparedness applications.

Open-source intelligence (OSINT): Refers to a broad array of information and sources that are generally available, including information obtained from the media (newspapers, radio, television, etc.), professional and academic records (papers, conferences, professional associations, etc.), and public data (government reports, demographics, hearings, speeches, etc.).

Privacy by design: The practice of building privacy-protecting features into security practices and technologies to preempt privacy concerns.

Profiling: Involves the reliance by authorities on generalizations about the suspiciousness or dangerousness of people who display certain traits, characteristics, or behaviors.

Prüm Convention: Provides law enforcement agencies in each of the EU Member States that are signatories direct online access to each other's law enforcement databases and enables them to exchange DNA, fingerprints, and other personal data.

Public-private partnership (PPP): A form of cooperation between public authorities and economic operators. The primary aims of this cooperation are to fund, construct, renovate, or operate an infrastructure, or the provision of a service.

Rapid Border Intervention Team (RABIT): Rapidly deployable EU border guards who are armed and patrol cross-country land borders, to assist in border control, particularly on the EU's southern coastlines. They are assigned by FRONTEX in crisis situations on the EU's external border.

Resilience: Enduring power for transformation, renewal, and recovery through the flux of interactions and flow of events; the ability of a substance or object to spring back into shape; elasticity; the capacity to recover quickly from difficulties; toughness; or the ability to withstand, adapt, and quickly recover from stresses and shocks.

Risk: Commonly defined as potential for an unwanted outcome resulting from an incident, event, or occurrence, as determined by its likelihood and the associated consequences. Any definition of risk is socially constructed and inseparable from the sense-making process that shapes the risk itself. The definition of risk requires acknowledging different perspectives, cultures, and interpretations.

Safety: Often briefly defined as prevention of industrial or natural accidents. More substantially, the state of being "safe" is defined as the condition of being protected against physical, social, material, psychological, or other types or consequences of failure, damage, error, accidents, harm or any other event that could be considered non-desirable. Safety can also be defined as the control of recognized hazards to achieve an acceptable level of risk.

Schengen Acquis: The Schengen Agreement and associated rules and regulations.

Schengen Agreement: Signed in 1985 by five of the then ten Member States of the then European Community, with implementation not seen before 1995 after the European Union had been founded, the Agreement led to the creation of Europe's borderless Schengen area of today. It proposed the gradual abolition of border checks at the signatories' common borders. Was supplemented in 1990 by the Schengen Convention, which later led to the abolition of EU internal border controls and a common EU visa policy.

Securitization: An extreme version of politicization that enables the use of extraordinary means in the name of security. It describes the political process and discussions through which societal phenomena become understood and addressed as security issues. Securitization does not "construct" risks that have no correlate to reality, but rather requires perceived insecurity that is sustained by highlighting threatening qualities of the phenomenon.

Security: Often briefly defined as protection from intentional attacks. More substantially, security is the state of being free from danger or threat, or the perception thereof. The definition often also includes the process by which this freedom is thought, in its broadest sense the safeguarding of a nation's, society's, organization's, etc., commonly acquired values.

Security by design: Take-up of security issues at the conception stage of new products/systems in order to reduce vulnerabilities. This includes strengthening complementarity and cooperation between civil and defense technologies. It requires comprehensive risk assessment to identify threats posed to systems, and then plan security into every stage.

Security paradox: (1) A situation where measures intended to be security-enhancing draw public attention to the presence of a threat that then is overestimated in the public's perception, unreasonably increasing feelings of insecurity and self-perceived vulnerability; (2) perception of security measures as unfair by those they affect, causing disagreement with a nation's security policies and facilitating radicalization and violence; and (3) see moral hazard.

Signals intelligence (SIGINT): Refers to electronic transmissions collected by ships, planes, ground sites, or satellites.

Societal security: In the first place comprises ethical aspects (acceptability), citizens' perceptions of (in)security, and citizens' acceptance of security interventions (such as policies) and of technological solutions to security problems. It further comprises citizen resilience and a whole-of-community approach.

Structured analytic techniques (SAT): A box of tools to help the analyst mitigate cognitive limitations and pitfalls.

Telemetry intelligence (TELINT): Sometimes used to indicate data relayed by weapons during tests.

Terrorism: Commonly defined as violent acts (or the threat of violent acts) intended to create fear (terror), perpetrated for a religious, political, or ideological goal, and which deliberately target or disregard the safety of non-combatants (e.g., neutral military personnel or civilians).

Threat: An act of coercion wherein an act is proposed to elicit a negative response. It often is a communicated intent to inflict harm or loss.

Unmanned aerial system (UAS): The term emphasizes the importance of other elements beyond an Unmanned Aerial Vehicle itself, including processing of the (surveillance) data that are collected.

Unmanned aerial vehicle (UAV): A powered, aerial vehicle that does not carry a human operator, uses aerodynamic forces to provide vehicle lift, can fly autonomously or be piloted remotely, can be expendable or recoverable, and can carry a payload.

Violent extremists: Individuals supporting or committing ideologically motivated violence aimed at reaching certain political goals.

Wassenar Arrangement: An international effort by 41 nations, including the major exporters of security technologies, to establish common, binding rules on exports of security technologies.

Whole-of-community approach: An approach to homeland and civil security that reinforces the fact that all of the resources available must be leveraged in preparing for, protecting against, responding to, recovering from, and mitigating against all hazards, and that the needs of the entire community in each of these areas must be collectively responded to.

Bibliography

9/11 Commission Report. 2014. http://www.9-11commission.gov/report/911Report.pdf

"A Secure Europe in a Better World. European Security Strategy." Brussels, December 12, 2003. http://www.consilium.europa.eu/uedocs/cmsUpload/78367.pdf

Agrawal, Mukta, and Casey Church. "Resilience Return on Investment—An Impossible Argument?" *Analytic Service, Inc.*, April 24, 2012. http://tisp.org/index.cfm?pk=download&pid=10261&id=12606%E2%80%8E

Allen, Thad. "The Future of Homeland Security: The Evolution of the Homeland Security Department's Roles and Missions." Testimony before the U.S. Senate Committee on Homeland Security and Government Affairs, July 22, 2012.

Alperen, Martin J. *Foundations of Homeland Security: Law and Policy.* Hoboken, NJ: Wiley, 2011.

Amass, Sandra F., Arun K. Bhunia, Alok R. Chaturvedi, Daniel R. Dolk, Srinivas Peeta, and Mikhail J. Atallah, eds. *The Science of Homeland Security.* West Lafayette, IN: Purdue University Press, 2006.

American Red Cross. "Web Users Increasingly Rely on Social Media to Seek Help in a Disaster." Published August 9, 2010. http://www.redcross.org/portal/site/en/menuitem.94aae335470e233f6cf911df43181aa0/?vgnextoid=6bb5a96d0a94a210VgnVCM10000089f0870aRCRD

American Society of Civil Engineers. *Guiding Principles for the Nation's Critical Infrastructure.* Reston, VA: ASCE Publications, 2009.

Archick, Kristin, coord. "European Approaches to Homeland Security and Counterterrorism." Congressional Research Services Report for Congress, RL33573. July 24, 2006. http://www.fas.org/sgp/crs/homesec/RL33573.pdf

Australian Government, Department of Immigration and Multicultural Indigenous Affairs. *Managing the Border: Immigration Compliance.* 2004–05 Edition. Canberra: Commonwealth of Australia, 2005.

Australian Government, Department of the Prime Minister and Cabinet. *Strong and Secure: A Strategy for Australia's National Security. Australian National Security Strategy.* January 23, 2013. http://www.nationalsecurity.gov.au

Australian Human Rights Commission. *Asylum Seekers and Refugees Guide.* Canberra: AHRC, n.d. https://www.humanrights.gov.au/asylum-seekers-and-refugees-guide#who

Bacastow, Todd S., and Dennis J. Bellafiore. "Redefining Geospatial Intelligence." *American Intelligence Journal* (Fall 2009): 38–40.

Baker, George W., and Dwight W. Chapman, eds. *Man and Society in Disaster.* New York: Basic Books, 1962.

Balzacq, Thierry, ed. *Securitization Theory: How Security Problems Emerge and Dissolve.* London and New York: Routledge, 2011.

Balzacq, Thierry, Stefano Guzzini, Michael C. Williams, Ole Wæver, and Heikki Patomäki. "Forum: What Kind of Theory—If Any—Is Securitization?" *International Relations.* First published on October 21, 2014 as doi: 10.1177/0047117814526606. http://ire.sagepub.com/content/early/2014/10/27/0047117814526606.full.pdf+html

Banchoff, Thomas. "Institutions, Inertia and European Union Research Policy." *Journal of Common Market Studies* 40 (2002): 1–21.

Barry, Tom. "Drones over the Homeland. How Politics, Money and Lack of Oversight Have Sparked Drone Proliferation, and What We Can Do." *International Policy Report* (Center for International Policy, Washington, D.C.), April 2013. http://www.ciponline.org/images/uploads/publications/IPR_Drones_over_Homeland_Final.pdf

Bart, Elias. "Pilotless Drones: Background and Considerations for Congress Regarding Unmanned Aircraft Operations in the National Airspace System." Congressional Research Service report, R42718, September 10, 2012. http://www.fas.org/sgp/crs/natsec/R42718.pdf

Barton, Allen H. *Communities in Disaster: A Sociological Analysis of Collective Stress Situations.* Garden City, NY: Doubleday, 1969.

BBC News. "Q&A: Schengen Agreement." March 4, 2013. http://www.bbc.com/news/world-europe-13194723

BBC News Europe. "More than 1,000 Arrested in Huge EU Crime Crackdown." Published September 24, 2014. http://www.bbc.com/news/world-europe-29344979

Beary, Brian. "The Drone Debate Growths in Europe." The European Institute, November 2012. http://www.europeaninstitute.org/index.php/162-european-affairs/ea-november-2012/1651-the-drone-debate-grows-in-europe

Becker, Ernest. *The Denial of Death.* New York: Free Press, 1997.

Beckman, James. *Comparative Legal Approaches to Homeland Security and Anti-Terrorism.* Aldershot, Hampshire: Ashgate, 2007.

Belgian Ministry of the Interior. "Video Surveillance in Public Spaces—Is It Really Efficient?" Study of the Katholieke Hogeschool Zuid-West-Vlaanderen, Expertisecentrum Maatschappelijke Veiligheid for the Belgian Ministry of the Interior, Flemish version, June 2013.

Bellavita, Christopher, and Ellen Gordon. "Changing Homeland Security: Teaching the Core." *Homeland Security Affairs* 2, no. 1 (2006): Article 1. http://www.hsaj.org/hsa/volII/iss1/art1

Bloom, Mia. *Dying to Kill: The Allure of Suicide Terror.* New York: Columbia University Press, 2007.

Boukalas, Christos. *Homeland Security, Its Law and Its State: A Design of Power for the 21st Century.* New York: Routledge, 2014.

Bourne, Mike. *Understanding Security.* Houndmills, Basingstoke: Palgrave Macmillan, 2014.

Brady, Hugo. "The Schengen Crisis in the Framework of the Arab Spring." *Panorama* (2012). http://www.cer.org.uk/sites/default/files/schengen_arab_Spring_Dec12_IEMED.pdf

Brainy Quotes. "Albert Einstein." Last accessed March 1, 2015. http://www.brainyquote.com/quotes/authors/a/albert_einstein_8.html

Breaking Defense. "Hybrid War." Published February 15, 2013. http://breakingdefense.com/tag/hybrid-war

Brimmer, Esther, ed. *Transforming Homeland Security: U.S. and European Approaches.* Washington, D.C.: Center for Transatlantic Relations, 2006.

Brooke-Holland, Louisa. "Unmanned Aerial Vehicles (Drones): An Introduction." House of Commons Library. April 25, 2013. http://www.parliament.uk/briefing-papers/SN06493.pdf

Bruneau, Michel, Stephanie E. Chang, Ronald T. Eguchi, George C. Lee, Thomas D. O'Rourke, and Andrei M. Reinhorn. "A Framework to Quantitatively Assess and Enhance the Seismic Resilience of Communications." *Earthquakes Spectra* 19 (2003): 733–752.

Bullock, Jane A., George D. Haddow, and Damon P. Coppola. *Introduction to Homeland Security: Principles of All-Hazards Risk Management,* 4th ed. New York: Elsevier, 2013.

Bunge, Mario. "Diagnosing Pseudoscience." In *Philosophy in Crisis. The Need for Reconstruction*, by Mario Bunge, 161–189. Amherst, NY: Prometheus Books, 2001.

Burgess, J. Peter, ed. *The Routledge Handbook of New Security Studies.* Milton Park: Routledge, 2010.

Business Continuity Institute. *4th Annual Survey: Supply Chain Resilience 2012.* November 2012. http://www.zurich.com/internet/main/sitecollectiondocuments/reports/supply-chain-resilience2012.pdf

Buzan, Barry. *People, States, and Fear: An Agenda for International Security Studies in the Post-Cold War Era.* Boulder, CO: Rienner, 1991.

———. *People, States, and Fear: The National Security Problem in International Relations.* Chapel Hill: The University of North Carolina Press, 1983.

———. *People, States & Fear: The National Security Problem in International Relations.* Boulder, CO: Rienner, 2009.

Buzan, Barry, and Lene Hansen. *The Evolution of International Security.* Cambridge: Cambridge University Press, 2009.

Byram, Daniel, ed. *Introduction to Homeland Security.* New York: Pearson Custom Publishing, 2009.

Cantore, Carlo Maria. "We're One, but We're Not the Same: Enhanced Cooperation and the Tension between Unity and Asymmetry in the EU." *Perspectives on Federalism* 3, no. 3 (2011). http://on-federalism.eu/attachments/103_download.pdf

Carrera, Sergio. *The EU Border Management Strategy; FRONTEX and the Challenges of Irregular Immigration in the Canary Islands.* CEPS Working Document, no. 261. Brussels: Center for European Policy Studies, 2007.

Carrera, Serigio, and Elspeth Guild. *Joint Operation RABIT 2010—FRONTEX Assistance to Greece's Border with Turkey: Revealing the Deficiencies of Europe's Dublin Asylum System*. Brussels: Center for European Policy Studies, 2010.

CBS News. "Napolitano: 'You Cannot Seal a Border.'" May 22, 2014. http://houston.cbslocal.com/2014/05/22/napolitano-you-cannot-seal-a-border

Center for Infrastructure Protection and Homeland Security. *The CIP Report*. Arlington, VA, George Mason University, November 2014.

Central Intelligence Agency. "The Intelligence Cycle." Last published March 23, 2013. https://www.cia.gov/kids-page/6-12th-grade/who-we-are-what-we-do/the-intelligence-cycle.html

———. *A Tradecraft Primer: Structured Analytic Techniques for Improving Intelligence Analysis*. Center for the Study of Intelligence, March 2009.

Centre for Strategy and Evaluation Services. "Ex-Post Evaluation of the Preparatory Action on Security Research (PASR)." January 2011. http://ec.europa.eu/enterprise/policies/security/files/doc/interim_evaluation_of_fp7_security_ex_post_pasr_final_report_en.pdf

CEUSS | Center for European Security Studies, Sigmund Freud University Vienna. "Roadmap Proposal for the Development of a Trans-disciplinary Security Research Paradigm." Deliverable 8.3, EU security research project FOCUS (Foresight Security Scenarios—Mapping Research to a Comprehensive Approach to Exogenous EU Roles), March 2013. http://www.focusproject.eu/documents/14976/63ec147d-c599-4d61-b2c7-be3d4590d3ea

Clarke, John L., ed. *Armies in Homeland Security: American and European Perspectives*. Washington, D.C.: National Defense University Press, 2006.

Collins, John W. Jr., Tsung Y. (Bill) Soo Hoo, Michael Krantz, and Richard Cosgrove. "Creating an Executive Doctorate in Civil Security in the United States." *Journal of Homeland Security and Emergency Management* 9, no. 2 (2012): Article 3.

Commission Decision of 3 February 2004 on the Implementation of the Preparatory Action on the Enhancement of the European Industrial Potential in the Field of Security Research (2004/213/EC). *Official Journal of the European Union,* L 67, March 5, 2004, 18–22. http://eur-lex.europa.eu/LexUriServ/LexUriServ.do?uri=OJ:L:2004:067:0018:0022:EN:PDF

Commission of the European Communities. "Commission Communications. Implementation of Preparatory Action on the Enhancement of the European Industrial Potential in the Field of Security Research: Towards a Programme to Advance European Security through Research and Technology." COM(2004)72. February 3, 2004. http://eur-lex.europa.eu/LexUriServ/LexUriServ.do?uri=COM:2004:0072:FIN:EN:PDF

———. "Communication from the Commission to the Council, the European Parliament, the European Economic and Social Committee and the Committee of the Regions: Security Research. The Next Steps." COM(2004) 590 final. September 7, 2004. http://eur-lex.europa.eu/LexUriServ/LexUriServ.do?uri=COM:2004:0590:FIN:EN:PDF

———. "Examining the Creation of a European Border Surveillance System (EUROSUR)." COM(2008) 68 final. Brussels, February 13, 2008. http://eur-lex.europa.eu/LexUriServ/LexUriServ.do?uri=COM:2008:0068:FIN:EN:PDF

Commission on the Intelligence Capabilities of the United States Regarding Weapons of Mass Destruction. *Report to the President of the United States*. Washington, D.C.: Government Printing Office, March 31, 2005.

"Comparative Study: IPR-Related Provisions in the US and EU R&D Funding Regimes." Unpublished study for the European Commission, 2014.

Consolidated Versions of the Treaty on European Union and the Treaty on the Functioning of the European Union. *Official Journal of the European Union,* C 326, October 26, 2012, 13–199. http://eur-lex.europa.eu/legal-content/EN/TXT/PDF/?uri=OJ:C:2012:326:FULL&from=EN

Cooper-Ramo, Joshua. *The Age of the Unthinkable: Why the New World Disorder Constantly Surprises Us.* New York: Back Bay Books, 2009.

Corry, Olaf. "Securitisation and 'Riskification:' Second Order Security and the Politics of Climate Change." *Millennium: Journal of International Studies* 40 (2011): 235–258.

Council of the European Union. "Draft Internal Security Strategy for the European Union: 'Towards a European Security Model.'" Brussels, March 8, 2010. http://register.consilium.europa.eu/pdf/en/10/st07/st07120.en10.pdf

———. *Prüm Convention.* July 7, 2005. http://ec.europa.eu/anti_fraud/documents/data-protection/dpo/prumtr.pdf

———. "Towards a European Strategy for the Development of Civil Applications of Remotely Piloted Aircraft Systems (RPAS)." 13438/12. Brussels, September 6, 2012. http://register.consilium.europa.eu/pdf/en/12/st13/st13438.en12.pdf

Council Regulation (EC) No. 428/2009 of 5 May 2009, Setting Up a Community Regime for the Control of Exports, Transfer, Brokering and Transit of Dual-Use Items (Recast). *Official Journal of the European Union,* L 134, May 29, 2009, 1–269. http://eur-lex.europa.eu/LexUriServ/LexUriServ.do?uri=OJ:L:2009:134:0001:0269:en:PDF

Covey, Stephen R. *The Seven Habits of Highly Effective People.* New York: Simon & Schuster, 1989.

Dalgaard-Nielsen, Anja, and Daniel S. Hamilton, eds. *Transatlantic Homeland Security: Protecting Society in the Age of Catastrophic Terrorism.* London: Routledge, 2006.

Decision No 1313/2013/EU of the European Parliament and of the Council of 17 December 2013 on a Union Civil Protection Mechanism. *Official Journal of the European Union,* L 347, December 20, 2013, 924–947. http://eur-lex.europa.eu/LexUriServ/LexUriServ.do?uri=OJ:L:2013:347:0924:0947:EN:PDF

DeGarmo, Matthew T. *Issues Concerning Integration of Unmanned Aerial Vehicles in Civil Airspace.* McLean, VA: MITRE Center for Advanced Aviation System Development, 2004. http://www.mitre.org/sites/default/files/pdf/04_1232.pdf

"Detection Technologies, Terrorism, Ethics, and Human Rights" (DETECTER). EU security research project. Last accessed March 6, 2015. http://www.detecter.eu

Deutsche Welle. "EU's 'Triton' Operation Gets Underway in Mediterranean." Published November 1, 2014. http://www.dw.de/eus-triton-operation-gets-underway-in-mediterranean/a-18033932

Dillon, Martin C. "Reversibility and Ethics." *Bulletin de la Société Américaine de Philosophie de Langue Française* 10, no. 2 (1998): 82–101.

Dorman, Andrew, and Joyce Kaufman, eds. *Providing for National Security: A Comparative Analysis.* Stanford, CA: Stanford University Press, 2014.

Dory, Amanda J. *Civil Security: Americans and the Challenge of Homeland Security.* Washington, D.C.: Center for Strategic and International Studies, 2003.

Douglas, Mary, and Aaron Wildavsky. *Risk and Culture: An Essay on the Selection of Technological and Environmental Dangers*. Berkeley and Los Angeles, CA: University of California Press, 1983.

Drones & Aerial Robotics Conference (DARC). New York University (NYU) School of Law, October 11–13, 2013. Last accessed March 19, 2015. http://droneconference.org

Drucker, Peter F. *The Essential Drucker*. New York: Harper Business, reissue edition, 2001.

Dynes, Russell R., and Kathleen J. Tierney, eds., *Disasters, Collective Behavior, and Social Organization*. Newark: University of Delaware Press, 1994.

Egli, Dane S. *Beyond the Storms: Strengthening Homeland Security & Disaster Management to Achieve Resilience*. Armonk, NY: M.E. Sharpe, 2014.

———, team lead. *Facing the Storms: Operationalizing Preparedness & Critical Infrastructure Resilience*. Laurel, MD: Johns Hopkins University Applied Physics Laboratory, 2013. http://www.jhuapl.edu/ourwork/nsa/papers/FacingtheStormsFinal-1-Oct-2013.pdf

Egli, Dane [S.], and Jared McKinney. "Economics of Resilience." In *The CIP Report*. By Center for Infrastructure Protection and Homeland Security, 6–8. Arlington, VA, George Mason University, November 2014. http://cip.gmu.edu/wp-content/uploads/2013/06/December2014_CIPHS_TheCIPReport_InvestmantsandAssessments.pdf

Erhard, Werner, Michael C. Jensen, and Kari L. Granger. "Creating Leaders: An Ontological/Phenomenological Model." In *The Handbook for Teaching Leadership*. Edited by Scott Snook, Nitin Nohria, and Rakesh Khurana, 245–262. Thousand Oaks, CA: Sage Publications, 2012.

Eriksson, E. Anders, Mattias Johansson, and Martin Lundmark. *Civil-Military Synergies in Research and Technology: Comparison of Sweden to Six European Nations*. Stockholm: FOI, 2014.

EUABC.com. "Enhanced Co-operation." Last accessed March 9, 2015. http://en.euabc.com/word/332

eu-LISA website. Last accessed March 7, 2015. https://www.eulisa.europa.eu

European Commission. "Communication from the Commission to the Council, the European Parliament, the European Economic and Social Committee and the Committee of the Regions: European Defence—Industrial and Market Issues. Towards an EU Defence Equipment Policy." COM(2003) 113(01). Brussels, March 11, 2003.

———. "European Commission Action Paper in Response to the Terrorist Attacks on Madrid." MEMO/04/66. March 18, 2004. http://europa.eu/rapid/press-release_MEMO-04-66_en.htm?locale=en

———. "European Commission Decision C (2014)4995 of 22 July 2014, Horizon 2020 Work Programme 2014–2015. 14. Secure Societies—Protecting Freedom and Security of Europe and Its Citizens. Revised." http://ec.europa.eu/research/participants/data/ref/h2020/wp/2014_2015/main/h2020-wp1415-security_en.pdf

———. "European Industry Leaders and EU Policymakers Meet to Plan for Security Research." Press release, IP/03/1351. Brussels, October 7, 2003. http://europa.eu/rapid/press-release_IP-03-1351_en.htm

———. "Guidance Notes on Project Reporting, FP7 Collaborative Projects, Networks of Excellence, Coordination and Support Actions, Research for the Benefit of Specific Groups (in particular SMEs). Version 2012." Last accessed March 9, 2015. http://ec.europa.eu/research/participants/data/ref/fp7/89692/project-reporting_en.pdf

———. "HORIZON 2020. The EU Framework Programme for Research and Innovation: Secure Societies—Protecting Freedom and Security of Europe and Its Citizens." Last accessed March 6, 2015. http://ec.europa.eu/programmes/horizon2020/en/h2020-section/secure-societies-%E2%80%93-protecting-freedom-and-security-europe-and-its-citizens

———. "Horizon 2020 Work Programme 2014–2015." European Commission Decision C (2014)4995, July 22, 2014, http://ec.europa.eu/research/participants/portal/doc/call/h2020/common/1617621-part_19_general_annexes_v.2.0_en.pdf

———. "More Research for Europe—Towards 3% of GDP." Commission Staff Working Paper in Support to the Commission Communication. COM(2002)499. September 11, 2002. http://ec.europa.eu/research/era/pdf/com3percent_en.pdf

———. "Questions and Answers: Schengen Information System (SIS II)." Press memo, April 9, 2013, http://europa.eu/rapid/press-release_MEMO-13-309_en.htm

———. "Special Eurobarometer 382: Public Attitudes towards Robots. Summary." September 2012. http://ec.europa.eu/public_opinion/archives/ebs/ebs_382_sum_en.pdf

———. "Taxation and Customs Union—The Customs Laboratories European Network (Overview)." Last published February 4, 2015. http://ec.europa.eu/taxation_customs/customs/customs_controls/customs_laboratories/group_ecl/index_en.htm

European Commission, Directorate General for Internal Policies, Policy Department; Citizens' Rights and Constitutional Affairs. *Review of Security Measures in the 7th Research Framework Programme: A Study*. 2010. http://www.statewatch.org/news/2010/nov/ep-review-security-research-programme.pdf

European Commission, Directorate General for Internal Policies, Policy Department; Citizens' Rights and Constitutional Affairs. *Review of Security Measures in the 7th Research Framework Programme: A Study*. 2014. http://www.europarl.europa.eu/RegData/etudes/etudes/join/2014/509979/IPOL-LIBE_ET(2014)509979_EN.pdf

European Commission, Enterprise and Industry Directorate-General. "Study Analysing the Current Activities in the Field of UAV. Second Element: Way Forward." ENTR/2007/065. http://ec.europa.eu/enterprise/policies/security/files/uav_study_element_2_en.pdf

European Commission, Humanitarian Aid and Civil Protection. "Emergency Response Coordination Centre (ERCC)." Last published June 10, 2014. http://ec.europa.eu/echo/en/what/civil-protection/emergency-response-coordination-centre-ercc

European Commission, Research and Innovation. "Research Ethics." Last published March 9, 2015. http://ec.europa.eu/research/swafs/index.cfm?pg=policy&lib=ethics

European Defence Agency. "Research and Technology Strategy." Last accessed March 9, 2015. https://eda.europa.eu/Aboutus/Whatwedo/strategies/ResearchandTechnology.

European Gendarmerie Force. "The European Gendarmerie Force." Last accessed March 7, 2015. http://www.eurogendfor.org/organization/what-is-eurogendfor

European Security Research Advisory Board (ESRAB). *Meeting the Challenge: The European Security Research Agenda. Report from the European Security Research Advisory Board.* Luxembourg: Office for Official Publications of the European Communities, 2006. http://ec.europa.eu/enterprise/policies/security/files/esrab_report_en.pdf

European Security Research and Innovation Forum (ESRIF). *ESRIF Final Report.* Brussels, 2009. http://ec.europa.eu/enterprise/policies/security/files/esrif_final_report_en.pdf

EUROPOL website. Last accessed March 7, 2015. https://www.europol.europa.eu

Executive Order 13526 of December 29, 2009, Classified National Security Information. *Code of Federal Regulations*, Title 3 (2010): 298–327. http://www.gpo.gov/fdsys/pkg/CFR-2010-title3-vol1/pdf/CFR-2010-title3-vol1-eo13526.pdf

Fabbrini, Federico. *The Enhanced Cooperation Procedure: A Study in Multispeed Integration.* N.p.: Centro Studi sul Federalismo, 2012. http://www.csfederalismo.it/attachments/2433_CSF-RP_Fabbrini_ENHANCED%20COOPERATION_October2012.pdf

Faber, Peter, and Valeri Ratchev, eds. *Homeland/Societal Vulnerability and Security.* Special issue, *Information & Security* 17 (2005). http://procon.bg/17-homeland-societal-vulnerability-and-security

Falkow, Michael D. "Does Homeland Security Constitute an Emerging Academic Discipline?" Master's thesis, Naval Postgraduate School, 2013. https://www.hsdl.org/?view&did=736325

Federal Aviation Administration. "Overview of Small UAS Notice of Proposed Rulemaking." Last published March 18, 2015. https://www.faa.gov/uas/nprm

———. "Press Release—DOT and FAA Propose New Rules for Small Unmanned Aircraft Systems." Published February 15, 2015. http://www.faa.gov/news/press_releases/news_story.cfm?newsId=18295

———. "Small UAS Notice of Proposed Rulemaking (NPRM)." Last accessed March 18, 2015. https://www.faa.gov/uas/nprm

———. "SWIM Program Overview." Last published February 6, 2015, http://www.faa.gov/nextgen/programs/swim/overview

Federal Bureau of Investigation. "Intelligence Collection Disciplines (INTs)." Last accessed February 23, 2015. http://www.fbi.gov/about-us/intelligence/disciplines

Federal Emergency Management Agency. "FY 2014 Tribal Homeland Security Grant Program (THSGP)." Last published July 25, 2014. https://www.fema.gov/fy-2014-tribal-homeland-security-grant-program-thsgp

———. *A Whole Community Approach to Emergency Management: Principles, Themes, and Pathways for Action.* FDOC-104-008-1. Washington, D.C., December 1, 2011. http://www.fema.gov/media-library-data/20130726-1813-25045-0649/whole_community_dec2011__2_.pdf

Federal Judicial Center. "Foreign Intelligence Surveillance Court." Last accessed February 23, 2014. http://www.fjc.gov/history/home.nsf/page/courts_special_fisc.html

Ferro, Carme, David Henry, and Thomas MacLellan. *A Governor's Guide to Homeland Security.* Washington, D.C.: National Governors Association (NGA), NGA Center for Best Practices, Homeland Security & Public Safety Division, November 2010. http://www.nga.org/files/live/sites/NGA/files/pdf/1011GOVGUIDEHS.PDF

Findlay, Keith A., and Michael S. Scott. "The Multiple Dimensions of Tunnel Vision in Criminal Cases." *Wisconsin Law Review* 2 (2006): 291–396.

Fischer, Robert J., Edward P. Halibozek, and David C. Walters. *Introduction to Security.* 9th ed. Waltham, MA: Butterworth-Heinemann, 2013.

Flammini, Francesco, Roberto Setola, and Giorgio Franceschetti, eds. *Effective Surveillance for Homeland Security: Balancing Technology and Social Issues.* Boca Raton, FL: CRC Press, 2013.

Flynn, Stephen. *The Edge of Disaster: Rebuilding a Resilient Nation.* New York: Random House, 2007.

Flynn, Stephen, and Sean Burke. *Critical Transportation Infrastructure and Societal Resilience.* Washington, D.C.: Center for National Policy, 2012.

Fosdick, Raymond B. "European Police Systems." *Journal of Criminal Law and Criminology* 6, no. 1 (1915): Article 3. http://scholarlycommons.law.northwestern.edu/cgi/viewcontent.cgi?article=1350&context=jclc

Franceschetti, Giorgio, and Marina Grossi, eds. *Homeland Security: Threats, Countermeasures, and Privacy Issues.* Norwood, MA: Artech House, 2011.

Freedberg, Sydney J. "Gen. Odierno: Budget Crunch Will Render Army Unready for Syria & Hybrid War." *Breaking Defense,* May 7, 2013. http://breakingdefense.com/2013/05/gen-odierno-budget-crunch-will-render-army-unready-for-syria-or-anywhere-else

FRONTEX website. Last accessed March 7, 2015. http://frontex.europa.eu

Fundamental Rights Agency of the European Union. *Access to Data Protection Remedies in EU Member States.* Luxembourg: Publications Office of the European Union, 2013. http://fra.europa.eu/sites/default/files/fra-2014-access-data-protection-remedies_en_0.pdf

Gartrell, Adam. "Teams to Give Verdict on Island Facilities." *Sydney Morning Herald,* April 20, 2012. http://www.smh.com.au/national/teams-to-give-verdict-on-island-facilities-20120819-24gr1.html

Gerencser, Mark, Reginald Van Lee, Fernando Napolitano, Christopher Kelly, and Walter Isaacson. *Megacommunities: How Leaders of Government, Business, and Non-Profits Can Tackle Today's Global Challenges Together.* New York: Palgrave Macmillan, 2008.

Gibbs, Jennifer. "The Relationship between Foreign Military Presence and Terrorist Attacks on Police." Paper presented at Academy of Criminal Justice Sciences, Philadelphia, PA, 2014.

Gill, Martin, ed. *The Handbook of Security.* 2nd ed. Basingstoke: Palgrave Macmillan, 2014.

Goldman, Alvin. "Social Epistemology." In *Stanford Encyclopedia of Philosophy.* Edited by Edward N. Zalta. Last published August 18, 2006. http://plato.stanford.edu/entries/epistemology-social

Goodman, H.A. "Illegal Immigrants Benefit the US Economy." *The Hill,* April 23, 2014. http://www.thehill.com/blogs/congress-blog/foreign-policy.203984-illegal-immigrants-benefit-the-us-economy

Gormley, Dennis M., and Richard Speier. "Controlling Unmanned Air Vehicles: New Challenges." *Nonproliferation Review* 10, no. 2 (2013). http://cns.miis.edu/npr/pdfs/102gorm.pdf

Gould, John B., Julia Carrano, Richard Leo, and Joseph Young. "Predicting Erroneous Convictions: A Social Science Approach to Miscarriages of Justice." Research report submitted to the U.S. Department of Justice, February 2013. https://www.ncjrs.gov/pdffiles1/nij/grants/241389.pdf

Guild, Elspeth. *Danger—Borders under Construction: Assessing the First Five Years of Border Policy in an Area of Freedom, Security, and Justice* (unpublished). Last accessed February 24, 2015. http://www.lse.ac.uk/europeanInstitute/research/hellenicObservatory/pdf/Various/ElspethPAPER.pdf

Gunderson, Lance H., and Crawford S. Holling, eds. *Panarchy: Understanding Transformations in Human and Natural Systems.* Washington, D.C.: Island Press, 2002.

Haddal, Chad C., and Jeremiah Gertler. "Homeland Security: Unmanned Aerial Vehicles and Border Surveillance." Congressional Research Service report, RS21698, July 8, 2010. http://fas.org/sgp/crs/homesec/RS21698.pdf

Hägerstrand, Torsten. "What about People in Regional Science?" *Papers of the Regional Science Association* 24, no. 1 (1970): 7–21.

Hansson, Sven Ove. "Science and Pseudo-Science." In *Stanford Encyclopedia of Philosophy.* Edited by Edward N. Zalta. Last published February 10, 2014. http://plato.stanford.edu/entries/pseudo-science, http://plato.stanford.edu/entries/pseudo-science

Harcourt, Bernard E. *Profiling, Punishing, and Policing in an Actuarial Age*. Chicago, IL: Chicago University Press, 2006.

Harel, Amos. "On Israel-Egypt Border, Best Defense Is a Good Fence." *Ha'aretz,* November 13, 2011. http://www.haaretz.com/print-edition/features/on-israel-egypt-border-best-defense-is-a-good-fence-1.395239

Harshbarger, Rebecca. "Terrorists Plotting to Blow Up 5 Planes in Christmas 'Spectacular': Report." *New York Post,* November 30, 2014. http://nypost.com/2014/11/30/terrorists-plotting-to-blow-up-5-planes-in-christmas-spectacular-report

Hart, David K. "The Moral Exemplar in an Organizational Society." In *Exemplary Public Administrators: Character and Leadership in Government.* Edited by Terry L. Cooper and N. Dale Wright, 9–29. San Francisco, CA: Jossey-Bass, 1992.

Harvey, Frank P. *The Homeland Security Dilemma: Fear, Failure and the Future of American Insecurity.* New York: Routledge, 2008.

Hayes, Ben. "Arming Big Brother. The EU's Security Research Programme." *Transnational Institute Briefing Series,* no. 1/2006. http://www.statewatch.org/news/2006/apr/bigbrother.pdf

———. *NeoConOpticon: The EU Security-Industrial Complex.* N.p.: Transnational Institute in association with Statewatch, 2009. http://www.statewatch.org/analyses/neoconopticon-report.pdf

Hendricks, Kevin B., and Vinod Singhal. "The Effect of Supply Chain Disruptions on Long-Term Shareholder Value, Profitability, and Share Price Volatility." *Supply Chain Magazine* (June 2005). http://www.supplychainmagazine.fr/TOUTE-INFO/ETUDES/singhal-scm-report.pdf

———. "An Empirical Analysis of the Effect of Supply Chain Disruptions on Long-Run Stock Price Performance and Equity Risk of the Firm." *Production and Operations Management* 4, no. 1 (2005): 35–52. http://www.wilfridlaurieruniversity.ca/documents/17398/scm_glitches_and_shareholder_risk.pdf

Heuer, Richards J. Jr. *Psychology of Intelligence Analysis.* Washington, D.C.: Center for the Study of Intelligence, 1999.

Hill, Jenny. "Germany Struggles to Adapt to Immigrant Influx." *BBC News,* November 2, 2014. http://www.bbc.com/news/world-europe-29686248

Hodson, Hal. "Pirates Incoming! Ship Radar Keeps Watch and Hits Back." *New Scientist,* May 30, 2014. http://www.newscientist.com/article/mg22229714.700-pirates-incoming-ship-radar-keeps-watch-and-hits-back.html#.U42qLUJidUQ

Home Office (UK). *Surveillance Camera Code of Practice.* London: The Stationery Office, 2013. https://www.gov.uk/government/uploads/system/uploads/attachment_data/file/204775/Surveillance_Camera_Code_of_Practice_WEB.pdf

Homeland Security News Wire. "CBP Flew Its Drones on Behalf of Other Agencies." Published January 22, 2014. http://www.homelandsecuritynewswire.com/dr20140122-cbp-flew-its-drones-on-behalf-of-other-agencies

———. "Infrastructure Protection: Drones to Help Assess Post-Disaster Infrastructure Damage." Published March 6, 2015. http://www.homelandsecuritynewswire.com/dr20150306-drones-to-help-assess-postdisaster-infrastructure-damage

Human Rights Watch. "They Know Everything We Do." N.p.: Human Rights Watch, 2014. http://www.hrw.org/reports/2014/03/25/they-know-everything-we-do

"Human Security Report Project." Last accessed March 29, 2015. http://www.hsrgroup.org

Immigrant Settlement Services of British Columbia. *Refugees: The Big Picture.* Vancouver: ISS of BC, n.d. Last accessed February 24, 2015. http://issbc.org/prim-corp-nav/our-work-with-refugees/Refugees-big-picture

Immigration Policy Center. *Refugees: A Fact Sheet.* Washington, D.C.: IPC, 2014. Last accessed February 24, 2015. http://www.immigrationpolicy.org/just-facts/refugees-fact-sheet

"Increasing Resilience in Surveillance Societies" (IRISS). EU security research project. Last accessed March 6, 2015. http://irissproject.eu

"Increasing Resilience in Surveillance Societies" (IRISS). EU security research project. Deliverable 1.1, "Surveillance, Fighting Crime and Violence," February 1, 2012. http://irissproject.eu/wp-content/uploads/2012/02/IRISS_D1_MASTER_DOCUMENT_17Dec20121.pdf

Industry Canada. *Canadian Industry Statistics.* Ottawa: Government of Canada, 2013. Last accessed February 24, 2015. http://www.ic.gc.ca/eic/site/cis-sic.nsf/eng/h_00029.html

Insurance Information Institute. "New App from Insurance Information Institute Helps Users Prepare for a Catastrophe before Disaster Strikes." Published August 14, 2012. http://www.iii.org/press-release/new-app-from-insurance-information-institute-helps-users-prepare-for-a-catastrophe-before-disaster-strikes-081412

Israel Ministry of Defense. *Israel's Security Fence Operational Concept.* Tel Aviv: MOD, January 2007.

Jackson, Susan, and Mihaly Csikszentmihalyi. *Flow in Sports: The Keys to Optimal Experiences and Performance.* Champaign, IL: Human Kinetics, 1999.

Jakubowicz, Yonatan. "Illegal Infiltration to Israel: It's the Economy Stupid!" *Jerusalem Post,* January 7, 2014. Last accessed: February 24, 2005. http://www.securityfence.mod.gov.il/Pages/ENG/operational.htm

Joh, Elizabeth E. "Policing by Numbers: Big Data and the Fourth Amendment." *Washington Law Review* 89, no. 35 (2014): 35–68.

Joint Chiefs of Staff. *Joint Publication 2–0, Joint Intelligence,* 23 October 2013. http://www.dtic.mil/doctrine/new_pubs/jp2_0.pdf

Joutsen, Matti. *The European Union and Cooperation in Criminal Matters: The Search for Balance.* HEUNI Paper, no. 25. Helsinki: The European Institute for Crime Prevention and Control, 2006.

Kahan, Jerome H. "What's in a Name? The Meaning of Homeland Security." *Journal of Homeland Security Education* 2 (2013): Article 2. http://www.journalhse.org/sft710/kahanjhsearticlefinal2.pdf

Kaunert, Christian, Sarah Leonard, and Patryk Pawlak, eds. *European Homeland Security: A European Strategy in the Making?* New York: Routledge, 2012.

Kellerman, Barbara. *The End of Leadership.* New York: HarperCollins Publishers, 2012.

Kenney, Michael. *From Pablo to Osama.* University Park, PA: Penn State University Press, 2008.

Kerry, John, Frank Lautenberg, and Kirsten Gillibrand. "Strengthening the Resiliency of Our Nation on the Ground (STRONG) Act (S-3691)." Introduced December 19, 2012. http://beta.congress.gov/bill/112th-congress/senate-bill/3691

Kettl, Donald. *System under Stress: Homeland Security and American Politics*. Washington, D.C.: CQ Press, 2004.

Kielkowski, Robin. "Leadership during Crisis." *Journal of Leadership Studies* 7, no. 3 (2013): 62–65.

Kierkegaard, Søren. *The Concept of Anxiety: A Simple Psychologically Orienting Deliberation on the Dogmatic Issue of Hereditary Sin*. Translated by Reidar Thomte and Albert B. Anderson. Princeton, NJ: Princeton University Press, 1981.

Kimmelmann, Jonathan. "The Promise and Peril of DNA Databanking." *Nature Biotechnology* 18 (2000): 695–696.

Kornhauser, William. *The Politics of Mass Society*. Glencoe, IL: Free Press, 1959.

Kowalski, Kathiann M. *A Pro/Con Look at Homeland Security: Safety vs. Liberty after 9/11*. Berkeley Heights, NJ: Enslow, 2008.

Kuhn, Thomas. *The Structure of Scientific Revolutions*. Chicago, IL: University of Chicago Press, 1962.

Lakatos, Imre, and Alan Musgrave, eds. *Criticism and the Growth of Knowledge*. Cambridge: Cambridge University Press, 1965.

Lakoff, George, and Mark Johnson. *Metaphors We Live By*. Chicago: University of Chicago Press, 2003.

Larsen, Jeffrey A., and Tasha L. Pravecek. *Comparative U.S.-Israeli Homeland Security*. Counterproliferation Paper, no. 34. Maxwell Air Force Base, AL: Air University, United States Air Force Counterproliferation Center, 2006. http://www.au.af.mil/au/awc/awcgate/cpc-pubs/larsen3.pdf

Lasswell, Harold D. *National Security and Individual Freedom*. New York: McGraw-Hill, 1950.

Lichfield, John. "EU Border Deal under Threat in Row Over Italy's Migrant Burden." *The Independent*, April 19, 2011. http://www.independent.co.uk/news/world/europe/eu-border-deal-under-threat-in-row-over-italys-migrant-burden-2269641.html

Liem, Khoen Daniel Hiller, and Christoph Castex. "A European Perspective on Security Research." In *European Perspectives on Security Research*. Edited by Klaus Thoma, 13–26. Berlin and Heidelberg: Springer, 2011. http://www.acatech.de/fileadmin/user_upload/Baumstruktur_nach_Website/Acatech/root/de/Publikationen/Englisch/acatech_diskutiert_European-Research_WEB.pdf

Lowenthal, Mark M. *Intelligence: From Secrets to Policy*. 6th edition. Los Angeles, CA: CQ Press, 2015.

Lubell, Maayan. "Israel Weed Smokers Choked by Border Fence with Egypt." *Reuters*, January 30, 2014. http://www.reuters.com/article/2014/01/30/us-israel-drugs-idUSBREA0T0MB20140130

Lyon, David. *The Electronic Eye: The Rise of the Surveillance Society*. Minneapolis: University of Minnesota Press, 1994.

Macnish, Kevin. "Unblinking Eyes: The Ethics of Automating Surveillance." *Ethics and Information Technology* 14, no. 2 (2012): 151–167.

Marchese, Kelly, Siva Paramasivam, and Michael Held. "Bouncing Back: Supply Chain Risk Management Lessons from Post-Tsunami Japan." *Industry Week*, March 9, 2012. http://www.

industryweek.com/global-economy/bouncing-back-supply-chain-risk-management-lessons-post-tsunami-japan

Martin, Gus. *Understanding Homeland Security*. Los Angeles, CA: Sage, 2015.

Marx, Gary T. "An Ethics for the New (and Old) Surveillance." In *Effective Surveillance for Homeland Security: Balancing Technology and Social Issues*. Edited by Francesco Flammini, Roberto Setola, and Giorgio Franceschetti, 3–19. Boca Raton, FL: CRC Press, 2013.

Masterman, Margret. "The Nature of a Paradigm." In *Criticism and the Growth of Knowledge*. Edited by Imre Lakatos and Alan Musgrave. Cambridge: Cambridge University Press, 1965, 59–90.

Maurer, Tim, Edin Omanovic, and Ben Wagner. "Uncontrolled Global Surveillance: Updating Export Controls for the Digital Age." New America Foundation, Open Technology Institute. March 2014. http://newamerica.net/sites/newamerica.net/files/policydocs/Uncontrolled_Surveillance_March_2014.pdf

Mawdsley, Jocelyn. "A European Agenda for Security Technology: From Innovation Policy to Export Controls." Flemish Peace Institute, report, January 2013. http://www.statewatch.org/news/2013/feb/fpi-european-agenda-security-technology.pdf

Maxwell, Bruce. *Homeland Security: A Documentary History*. Washington, D.C.: CQ Press, 2004.

McEntire, David A. "The Importance of Multi- and Interdisciplinary Research on Disasters and for Emergency Management." In *Disciplines, Disasters and Emergency Management: The Convergence and Divergence of Concepts, Issues and Trends from the Research Literature*. Edited by David A. McEntire, 3–14. Springfield, IL: Charles C. Thomas, 2007 [also available in an online version: Emmitsburg, MD: FEMA Emergency Management Institute. https://training.fema.gov/hiedu/aemrc/booksdownload/ddemtextbook]

———, ed. *Comparative Emergency Management: Understanding Disaster Policies, Organizations, and Initiatives from Around the World*. Emmitsburg, MD: FEMA Emergency Management Institute, 2013. https://training.fema.gov/hiedu/aemrc/booksdownload/compemmgmtbookproject

McIntire, Mike. "Ensnared by Error on Growing US Watch List." *New York Times*, April 6, 2010. http://www.nytimes.com/2010/04/07/us/07watch.html?pagewanted=all&_r=0

McKenzie, Jaffa, and Reza Hasmath. "Deterring the 'Boat People': Explaining the Australian Government's People Swap Response to Asylum Seekers." *Australian Journal of Political Science* 48, no. 4 (2013): 417–430.

Miller, Gabriel. "Activity-Based Intelligence Uses Metadata to Map Adversary Networks." *Defense News*, July 8, 2013. Last accessed February 23, 2015. http://www.defensenews.com/article/20130708/C4ISR02/307010020/Activity-Based-Intelligence-Uses-Metadata-Map-Adversary-Networks

Miller, Harvey J. "A Measurement Theory for Time Geography." *Geographical Analysis* 37 (2005): 17–45. http://onlinelibrary.wiley.com/doi/10.1111/j.1538-4632.2005.00575.x/full

Ministry of Defence. "Joint Doctrine Note 2/11: The UK Approach to Unmanned Aircraft Systems." March 30, 2011. https://www.gov.uk/government/uploads/system/uploads/attachment_data/file/33711/20110505JDN_211_UAS_v2U.pdf

Miskin, Maayana. "August: Zero Illegal Infiltrations along Israel-Egypt Border." *Arutz Sheva*, September 1, 2013, http://www.israelnationalnews.com/News/News.aspx/171488#.VKh2LyvF98E

Moeckli, Daniel. "Discriminatory Profiles: Law Enforcement after 9/11 and 7/7." *European Human Rights Law Review* 5 (2005): 517–532.

Monar, Jorg. "The Project of a European Border Guard: Origins, Models and Prospects in the Context of the EU's Integrated External Border Management." In *Borders and Security Governance*. Edited by Marina Caparini and Otwin Marenin. Geneva: Geneva Centre for the Democratic Control of Armed Forces, 2006.

Morag, Nadav. *Comparative Homeland Security: Global Lessons*. Hoboken, NJ: Wiley, 2011.

Morton, John, and Catherine Feinman. *DomPrep Action Plan: Building Resilient Regions for a Secure and Resilient Nation*. Severna Park, MD: IMR Group, Inc., 2012.

Moteff, John. *Critical Infrastructure Resilience: The Evolution of Policy and Programs and Issues for Congress*. Washington, D.C.: Congressional Research Service, August 23, 2012.

Mussington, David. "Concepts for Enhancing Critical Infrastructure Protection: Relating Y2K to CIP Research and Development." *Monograph Report*, RAND Corporation Science & Technology, 2002.

National Academies, Committee on Science, Engineering, and Public Policy, Committee on Increasing National Resilience to Hazards and Disasters. *Disaster Resilience: A National Imperative*. Washington, D.C.: The National Academies Press, 2012.

National Academy of Engineering. "Prevent Nuclear Terror." Last accessed February 25, 2015. http://www.engineeringchallenges.org/cms/8996/9134.aspx

National Geospatial-Intelligence Agency. *National System for Geospatial Intelligence, Geospatial Intelligence (GEOINT) Basic Doctrine Publication 1–0*. Office of Geospatial-Intelligence Management, September 2006.

National Infrastructure Advisory Council. *Best Practices for Government to Enhance the Security of National Critical Infrastructures*. April 13, 2004. http://www.dhs.gov/xlibrary/assets/niac/NIAC_BestPracticesSecurityInfrastructures_0404.pdf

———. *Cross Sector Interdependencies and Risk Assessment Guidance*, January 13, 2004. http://www.dhs.gov/xlibrary/assets/irawgreport.pdf

———. *Regional Resilience Scoping Study*, March 23, 2012. http://www.dhs.gov/xlibrary/assets/niac/niac-regional-resilience-white-paper-032412.pdf

National Legislative Bodies/National Authorities. *Regional Resettlement Arrangement Between Australia and Papua New Guinea*. July 19, 2013. http://www.refworld.org/docid/51f61a504.html

National Research Council of the National Academies, Committee on Educational Paradigms for Homeland Security. *Frameworks for Higher Education in Homeland Security*. Washington, D.C., 2005.

Neal, Andrew W. "Securitizing and Risk at the EU Border: The Origins of FRONTEX." *Journal of Common Market Studies* 47 (2009): 333–356.

Nelson, Rick "Ozzie," project director. *U.S.-India Homeland Security Cooperation: Building a Lasting Partnership via Transportation Sector Security*. Center for Strategic and International Studies: A Report for the CSIS Homeland Security and Counterterrorism Program. Lanham, MD: Rowman & Littlefield, 2013.

Neubauer, Ian L. "Australia's Plan to Outsource Its Refugee Problem to Cambodia Won't Work." *Time*, September 29, 2014. http://www.time.com/3443252/australia-asylum-seeker-refugee-cambodia

Nielsen, Nikolaj. "Ashton Calls for Military-Grade Drones in EU Airspace." *EUobserver,* October 22, 2013. http://euobserver.com/defence/121854

Nwosu, Chiamaka, Jeanne Batalova, and Gregory Auclair. *Frequently Requested Statistics on Immigrants and Immigration in the United States.* Washington, D.C.: Migration Policy Institute, 2014. http://www.migrationpolicy.org/article/frequently-requested-statistics-immigrants-and-immigration-united-states

O'Connor, Michael P., and Celia M. Rumann. "Into the Fire: How to Avoid Getting Burned by the Same Mistakes Made Fighting Terrorism in Northern Ireland." *Cardozo Law Review* 24 (2003): 1657–1751.

O'Rourke, Thomas D. "Critical Infrastructure, Interdependencies, and Resilience." *The Bridge. Linking Engineering and Society* 37, no. 1 (2007). https://www.nae.edu/File.aspx?id=7405

Office of the Director of National Intelligence. *Intelligence Community Directive 203, Analytic Standards,* January 2, 2015.

———. "Members of the IC." Last accessed February 23, 2013. http://www.dni.gov/index.php/intelligence-community/members-of-the-ic

Okun, Barbara F. *Seeking Connections in Psychotherapy.* Hoboken, NJ: Jossey-Bass, 1990.

Pape, Robert A. *Dying to Win.* New York: Random House, 2005.

Paquin, Jonathan, and Patrick James, eds. *Game Changer: The Impact of 9/11 on North American Security.* Vancouver and Toronto: UBC Press, 2014.

Pelfrey, William V., and William D. Kelley. "Homeland Security Education: A Way Forward." *Homeland Security Affairs* 9 (2013): Article 3. https://www.hsaj.org/articles/235

Penn State Homeland Security Programs. "TMI@35: The Three Mile Island Nuclear Crisis in Perspective. Analyses, Stories, Policies. Documentation of a Conference Held at Penn State Harrisburg, March 27–28, 2014." Last accessed March 29, 2015. http://sites.psu.edu/tmi35

Penn State World Campus. "Master of Professional Studies in Homeland Security." Last accessed March 29, 2015. http://www.worldcampus.psu.edu/degrees-and-certificates/homeland-security/overview

Peoples, Columba, and Nick Vaughan-Williams. *Critical Security Studies: An Introduction,* 2nd ed. London and New York: Routledge, 2015.

Phillips, Leigh. "EU 'Homeland Security' Lacks Democratic Oversight, Says Watchdog." *EUobserver,* October 2, 2009. https://euobserver.com/justice/28759

Piaguet, Carine, and Frédéric Gouardères. "Fact Sheets on the European Union—Defence Industry." European Parliament, March 2014. http://www.europarl.europa.eu/aboutparliament/en/displayFtu.html?ftuId=FTU_5.9.5.html

Plant, Jeremy, and Michael Kenney. "Center for Homeland Security and Terrorism Studies." Concept paper, Penn State Harrisburg, School of Public Affairs, 2004.

Police Department under the Ministry of the Interior of the Republic of Lithuania, Presidency of the Council of the European Union. "European Network of Law Enforcement Technology Services." Last published July 16, 2013. http://www.policija.lt/index.php?id=21280

Popper, Karl. *The Open Society and Its Enemies.* London: Routledge, 1945.

Porter, Kay. *The Mental Athlete: Inner Training for Peak Performance in All Sports.* Champaign, IL: Human Kinetics, 2003.

Presidential Policy Directive 8 (PPD-8), National Preparedness. March 30, 2011. http://www.dhs.gov/presidential-policy-directive-8-national-preparedness

Presidential Policy Directive 21 (PPD-21), Critical Infrastructure Security and Resilience. February 12, 2013. http://www.whitehouse.gov/the-press-office/2013/02/12/presidential-policy-directive-critical-infrastructure-security-and-resil

Privacy International. "The Surveillance Industry Index." November 2013. Last published May 23, 2014. https://www.privacyinternational.org/?q=node/403

"Protecting Freedom and Security of Europe and Its Citizens. European Commission Decision C (2014)4995 of 22 July 2014. Revised." http://ec.europa.eu/research/participants/data/ref/h2020/wp/2014_2015/main/h2020-wp1415-security_en.pdf

Rachels, James. "Why Privacy Is Important." *Philosophy and Public Affairs* 4 (1975): 323–333.

Ramsay, James, and Linda Kiltz. "Introduction." In *Critical Issues in Homeland Security: A Casebook.* Edited by James D. Ramsay and Linda Kiltz, 1–3. Boulder, CO: Westview, 2014.

Reese, Shaun. "Defining Homeland Security: Analysis and Congressional Considerations." Congressional Research Service, CRS Report for Congress (R42462), 2013. http://fas.org/sgp/crs/homesec/R42462.pdf

Regulation (EC) No. 562/2006 of the European Parliament and of the Council of 15 March 2006 Establishing a Community Code on the Rules Governing the Movement of Persons across Borders (Schengen Borders Code). *Official Journal of the European Union,* L 105, April 13, 2006, 1–32. http://eur-lex.europa.eu/LexUriServ/LexUriServ.do?uri=OJ:L:2006:105:0001:0032:EN:PDF

Regulation (EU) No. 1291/2013 of the European Parliament and of the Council of 11 December 2013 Establishing Horizon 2020—The Framework Programme for Research and Innovation (2014–2020) and Repealing Decision No. 1982/2006/EC. *Official Journal of the European Union,* L 347, December 20, 2013, 104–173. http://inea.ec.europa.eu/download/legal_framework/regulation_12912013_establishing_h2020.pdf

Research for a Secure Europe. Report of the Group of Personalities in the Field of Security Research. Luxembourg: Office for Official Publications of the European Communities, 2004. http://ec.europa.eu/enterprise/policies/security/files/doc/gop_en.pdf

Roberts, John. "EXCLUSIVE: Drones Vulnerable to Terrorist Hijacking, Researchers Say." *Fox News,* June 25, 2012. http://www.foxnews.com/tech/2012/06/25/drones-vulnerable-to-terrorist-hijacking-researchers-say

Roma, Alfredo. "European Commission Unmanned Aircraft System Panel Process: 4th Workshop on Societal Dimension." November 16, 2011. http://ec.europa.eu/enterprise/docs/uas/Report_Workshop_4.pdf

Rosenzweig, Ido, and Yuval Shany. "A Decade of Palestinian Terrorism—Report by the Israeli Security Agency." *Terrorism and Democracy,* no. 14 (n.d.). Last accessed February 24, 2015. http://www.en.idi.org.il/analysis/terrorism-and-democracy/issue-no-14/a-decade-of-palestinian-terrorism-%E2%80%93-report-by-the-israeli-security-agency

"Rules, Expectations & Security through Privacy-Enhanced Convenient Technologies" (RESPECT). EU security research project. Last accessed March 6, 2015. http://www.respectproject.eu

Salter, Mark B., and Can E. Mutlu, eds. *Research Methods in Critical Security Studies: An Introduction.* New York and London: Routledge, 2013.

Sauer, Frank, and Niklas Schörnig. "Killer Drones: The 'Silver Bullet' of Democratic Warfare?" *Security Dialogue* 43 (2012): 363–380.

Schneider, Kirk J. *The Paradoxical Self: Toward an Understanding of Our Contradictory Nature.* New York: Humanity Books, 1999.

Selk, Jason. *10-Minute Toughness: The Mental Training Program for Winning before the Game Begins.* New York: McGraw-Hill, 2008.

———. *Executive Toughness: The Mental-Training Program to Increase Your Leadership Performance.* New York: McGraw-Hill, 2011.

Shepherd, Laura J., ed. *Critical Approaches to Security: An Introduction to Theories and Methods.* New York and London: Routledge, 2013.

Silberglitt, Richard, Philip S. Antón, David R. Howell, and Anny Wong. *The Global Technology Revolution 2020: In-Depth Analyses. Bio/Nano/Materials/Information Trends, Drivers, Barriers, and Social Implications.* Santa Monica, CA: RAND, 2006. http://www.rand.org/content/dam/rand/pubs/technical_reports/2006/RAND_TR303.pdf

Skrzypietz, Therese. "Unmanned Aircraft Systems for Civilian Missions." *BIGS Policy Paper* (Brandenburg Institute for Society and Security), no. 1 (February 2012). http://www.bigs-potsdam.org/images/Policy%20Paper/PolicyPaper-No.1_Civil-Use-of-UAS_Druckversion.pdf

Smith, Anthony D. "States and Homelands: The Social and Geopolitical Implications of National Territory." *Millennium: Journal of International Studies* 10, no. 3 (1981): 187–202.

Smith, Clifton L., and David J. Brooks. *Security Science: The Theory and Practice of Security.* New York: Elsevier, 2013.

Smith, Robert W. "What Is Homeland Security?" *Journal of Public Affairs Education* 11, no. 3 (2005): 233–246.

Societal Impact Expert Working Group. "Report of the Societal Impact Expert Working Group." EC DG ENTR report, 2012. http://cies.ie/wp-content/uploads/2014/05/Report-of-the-Societal-Impact-Expert-Working-Group.pdf

Sorokin, Pitrim A. *Man and Society in Calamity: The Effects of War, Revolution, Famine, Pestilence upon Human Mind, Behavior, Social Organization and Cultural Life.* New York: Dutton, 1942.

Sparrow, Robert. "Building a Better War Bot: Ethical Issues in the Design of Unmanned Systems for Military Applications." *Science and Engineering Ethics* 15, no. 2 (2009): 169–187.

Stanley, Jay, and Catherine Crum. "Protecting Privacy from Aerial Surveillance: Recommendations for Government Use of Drone Aircraft." American Civil Liberties Union (ACLU), report, December 2011. http://www.aclu.org/files/assets/protectingprivacyfromaerialsurveillance.pdf

Steiner, James E. *Homeland Security Intelligence.* Thousand Oaks, CA: CQ Press, 2015.

"Surveillance: Ethical Issues, Legal Limitations, and Efficiency" (SURVEILLE). EU security research project. Last accessed March 6, 2015. http://www.surveille.eu

Texas Christian University, Neeley School of Business. "Weathering the Storm: How to Achieve Strategic Resilience through Supply Chain Excellence." Published June 15, 2012. http://www.neeley.tcu.edu/News_and_Events/Press_Releases/Weathering_the_Storm__How_to_Achieve_Strategic_Resilience_Through_Supply_Chain_Excellence.aspx

The Local FR. "France Rejects Four out of Five Asylum Seekers." March 24, 2014. http://www.thelocal.fr/20140324/france-asylum-refugee-rejects-four-out-of-five-asylum-seekers

The President of the United States. *National Security Strategy.* Washington, D.C.: The White House, February 2015. https://www.whitehouse.gov/sites/default/files/docs/2015_national_security_strategy.pdf

The President of the United States, Homeland Security Council. *National Strategy for Homeland Security.* Washington, D.C.: The White House, October 2007. http://www.dhs.gov/xlibrary/assets/nat_strat_homelandsecurity_2007.pdf

Thoma, Klaus, ed. *European Perspectives on Security Research.* Munich: acatech—Deutsche Akademie der Technikwissenschaften, 2011. http://www.acatech.de/fileadmin/user_upload/Baumstruktur_nach_Website/Acatech/root/de/Publikationen/Englisch/acatech_diskutiert_European-Research_WEB.pdf

Thompson, Richard M. III. "Drones in Domestic Surveillance Operations: Fourth Amendment Implications and Legislative Responses." Congressional Research Service report, R42701, April 3, 2013. http://fas.org/sgp/crs/natsec/R42701.pdf

Tow, William T., David Walton, and Rikki Kersten, eds. *New Approaches to Human Security in the Asia-Pacific: China, Japan and Australia.* Farnham, Surrey: Ashgate, 2013.

Travis, Alan. "Big Brother Traffic Cameras Must Be Regulated, Says Home Secretary." *The Guardian,* July 4, 2010. http://www.theguardian.com/uk/2010/jul/04/anpr-surveillance-numberplate-recognition

U.S. Code (1952) Title 35, Section 202—Disposition of Rights. http://www.gpo.gov/fdsys/pkg/USCODE-2011-title35/pdf/USCODE-2011-title35-partII-chap18-sec202.pdf

U.S. Code (1952) Title 35, Section 203—March-in Rights. http://www.gpo.gov/fdsys/pkg/USCODE-2011-title35/pdf/USCODE-2011-title35-partII-chap18-sec203.pdf

U.S. Code (2006), Supplement 5, Title 50—WAR AND NATIONAL DEFENSE, section 401a—Definitions. http://www.gpo.gov/fdsys/granule/USCODE-2011-title50/USCODE-2011-title50-chap15-sec401a

U.S. Code (2006), Title 10—ARMED FORCES, section 467. http://www.gpo.gov/fdsys/pkg/USCODE-2006-title10/html/USCODE-2006-title10-subtitleA-partI-chap22-subchapIV-sec467.htm

U.S. Customs and Border Protection. "About CBP." Last accessed March 29, 2015. http://www.cbp.gov/about

U.S. Department of Homeland Security. *The 2014 Quadrennial Homeland Security Review.* Washington, D.C., June 18, 2015. http://www.dhs.gov/sites/default/files/publications/2014-qhsr-final-508.pdf

———. "Countering Violent Extremism." Last published October 15, 2014. http://www.dhs.gov/topic/countering-violent-extremism.

———. *National Infrastructure Protection Plan.* 2009. http://www.dhs.gov/xlibrary/assets/NIPP_Plan.pdf

———. *National Preparedness Goal.* 1st ed., September 2011. http://www.fema.gov/media-library-data/20130726-1828-25045-9470/national_preparedness_goal_2011.pdf

———. *National Response Framework.* 2008. http://www.dhs.gov/xlibrary/assets/NRF.pdf

———. "National Terrorism Advisory System." Last published September 5, 2014. http://www.dhs.gov/national-terrorism-advisory-system

———. *Tribal Resource Guide.* Last published June 2013. http://www.dhs.gov/xlibrary/assets/iga/dhs-tribal-resource-guide.pdf

U.S. Department of Homeland Security, Office of the Inspector General (OIG). "Challenges in FEMA's Flood Map Modernization Program." Report no. OIG-05-44. September 2005.

U.S. Department of Transportation, Federal Aviation Administration. "Operation and Certification of Small Unmanned Aircraft Systems." Billing Code 4910-13-P, RIN 2120-AJ60. February 15, 2015. https://www.faa.gov/regulations_policies/rulemaking/recently_published/media/2120-AJ60_NPRM_2-15-2015_joint_signature.pdf

U.S. Department of Treasury. "A New Economic Analysis of Infrastructure Investment: A Report Prepared by the Department of Treasury with the Council of Economic Advisors." March 23, 2012. http://www.treasury.gov/resource-center/economic-policy/Documents/20120323InfrastructureReport.pdf

U.S. Government Accountability Office. *Critical Infrastructure: Challenges Remain in Key Sectors.* Testimony before the Subcommittee on Homeland Security, Committee on Appropriations, House of Representatives. GAO-07-626T. March 2007.

———. *Critical Infrastructure Protection: Update to National Infrastructure Protection Plan Includes Increased Emphasis on Risk Management and Resilience.* Report to Members and Committees of Congress. GAO-10-296. March 2010.

U.S. Resilience Project. "Business Case for Supply Chain Security and Resilience." March 1, 2012. http://www.usresilienceproject.org/workshop/participants/pdfs/USRP_Resources_Chapter_6_030112.pdf

"U.S. Supports Unarmed Domestic Drone." Monmouth University Poll. August 15, 2013. http://www.monmouth.edu/assets/0/32212254770/32212254991/32212254992/32212254994/32212254995/30064771087/409aecfb-3897-4360-8a05-03838ba69e46.pdf

"UAV-Based Innovative Means for Land and Sea Non-Cooperative Vehicles Stop" (AEROCEPTOR). EU security research project. Last accessed March 11, 2015. http://www.aeroceptor.eu

United Nations High Commissioner for Refugees. *Convention and Protocol Relating to the Status of Refugees.* Geneva: UNHCR, 1951. http://www.unhcr.org/protect/PROTECTION/3b66c2aa10.pdf

United States Census Bureau. *Top Trading Partners—December 2013.* Washington, D.C.: U.S. Census Bureau, 2013. https://www.census.gov/foreign-trade/statistics/highlights/top/top1312yr.html

Uniting and Strengthening America by Providing Appropriate Tools Required to Intercept and Obstruct Terrorism (USA PATRIOT) Act of 2001. http://www.gpo.gov/fdsys/pkg/BILLS-107hr3162enr/pdf/BILLS-107hr3162enr.pdf

Urban Securipedia. "Crisis Management Cycle." Last published August 15, 2013. http://securipedia.eu/mediawiki/index.php/Crisis_management_cycle

van Ewijk, Anne R. "Diversity within Police Forces in Europe: A Case for the Comprehensive View." *Policing* 5 (2011): 1–17.

Van Manen, Max. *Researching Lived Experience: Human Science for an Action Sensitive Pedagogy.* New York: State University of New York Press, 1990.

Van Wart, Montgomery, and Naim Kapucu. "Crisis Management Competencies: The Case of Emergency Managers in the USA." *Public Management Review* 13 (2011): 489–511.

Vasquez, Madeleine, Jerome Kisielewicz, Nihar Shembavnekar, Salvatore Petronella, James Brassington, and Mathieu Capdevila. "Study on the Feasibility of Improved Co-operation between Bodies Carrying out European Coast Guard Functions." ICF International, Final Report, June 2014. http://ec.europa.eu/transport/modes/maritime/studies/doc/2014-06-icf-coastguard.pdf

Vermeulen, Gert, and Wendy De Bondt. *Justice, Home Affairs and Security: European and International Institutional and Policy Development.* Antwerpen: Maklu, 2015.

Voeller, John G., ed. *Wiley Handbook of Science and Technology for Homeland Security.* Hoboken, NJ: Wiley, 2010.

Voisin, Gabriel. "Drones: Privacy Implications across the EU." Bird & Bird, July 15, 2013. http://www.twobirds.com/en/news/articles/2013/global/drones-privacy-implications-across-the-eu

Wakefield, Jane. "Huge Raid to Shut Down 400-Plus Dark Net Sites." *BBC News Technology.* Published November 7, 2014. http://www.bbc.com/news/technology-29950946

Wallace, John A. "Integrating Unmanned Aircraft Systems into Modern Policing in an Urban Environment." Thesis, Naval Postgraduate School, Monterey, CA, 2012. http://calhoun.nps.edu/public/bitstream/handle/10945/17474/12Sep_Wallace_John.pdf?sequence=1

Wassenaar Arrangement on Export Controls for Conventional Arms and Dual-Use Goods and Technologies, Secretariat. "Wassenaar Arrangement on Export Controls for Conventional Arms and Dual-Use Goods and Technologies." Last published February 12, 2015. http://www.wassenaar.org

Watson, James B. "Psychology as the Behaviorist Views It." *Psychological Review* 20 (1913): 158–177.

Weaver, Leon H. *The Civil Defense Debate: Differing Perceptions of a Persistent Issue in National Security Policy.* East Lansing, MI: Michigan State University, Social Science Research Bureau, 1967.

Whitehouse, Kaja. "Tech Crackdown on Drones Goes Local." *USA Today,* February 23, 2015. http://www.usatoday.com/story/tech/2015/02/23/crackdown-drones-technology-faa-nyc-councilman/23696377

Wikipedia. "Civil Defense." Last published March 14, 2015. http://en.wikipedia.org/wiki/Civil_defense

———. "Common European Security and Defence Policy." Last published January 2, 2015. http://en.wikipedia.org/wiki/Common_Security_and_Defence_Policy

———. "Community Acquis." Last published December 16, 2014. http://en.wikipedia.org/wiki/Community_acquis

———. "Defense Support of Civil Authorities." Last published June 20, 2014. http://en.wikipedia.org/wiki/Defense_Support_of_Civil_authorities; see also the *Homeland Defense and Civil Support Journal* published by the U.S. Army War College, http://www.csl.army.mil/homelanddefensejournal/JournalHome.aspx

———. "Delian League." Last published March 24, 2015. http://en.wikipedia.org/wiki/Delian_League

———. "Enhanced Cooperation." Last published February 12, 2015. http://en.wikipedia.org/wiki/Enhanced_cooperation

———. "German Wings Flight 9525." Last published March 28, 2015. http://en.wikipedia.org/wiki/Germanwings_Flight_9525

———. "Human Security." Last published March 28, 2015. http://en.wikipedia.org/wiki/Human_security

———. "Privacy by Design." Last published March 4, 2015. http://en.wikipedia.org/wiki/Privacy_by_Design

Wildavsky, Aaron. *Searching for Safety.* New Brunswick, NJ: Transaction Books, 1988.

Wirtz, James J. "The Sources and Methods of Intelligence Studies." In *The Oxford Handbook of National Security Intelligence.* Edited by Loch K. Johnson, 59–69. New York: Oxford University Press, 2010.

Wolfers, Arnold. "'National Security' as an Ambiguous Symbol." *Political Science Quarterly* 67 (1952): 481–502.

World Economic Forum. *Building Resilience in Supply Chains.* January 2013. http://www3.weforum.org/docs/WEF_RRN_MO_BuildingResilienceSupplyChains_Report_2013.pdf

Wuelser, Gabriela, Christian Pohl, and Gabriela Hirsch Hadorn. "Structuring Complexity for Tailoring Research Contributions to Sustainable Development: A Framework." *Sustainability Science* 7 (2011): 81–93.

Wuthnow, Robert. *Be Very Afraid: The Cultural Response to Terror, Pandemics, Environmental Devastation, Nuclear Annihilation, and Other Threats.* Oxford: Oxford University Press, 2010.

Yu, Yijun. "Germanwings Flight 4U9525: A Victim of the Deadlock between Safety and Security Demands." *Homeland Security News Wire,* March 30, 2015. http://www.homelandsecuritynewswire.com/dr20150330-germanwings-flight-4u9525-a-victim-of-the-deadlock-between-safety-and-security-demands?page=0,0

Zedner, Lucia. *Security.* London: Routledge, 2009.

Zolli, Andrew, and Ann Marie Healy. *Resilience: Why Things Bounce Back.* New York: Free Press, 2012.

Homeland and Civil Security Learning and Research Guide

Glossaries and Knowledge Bases

- Guide to Emergency Management and Related Terms, Definitions, Acronyms, Programs and Legislation. A Tutorial on Emergency Management, Broadly Defined, Past and Present (FEMA Emergency Management Institute, 2008). http://training.fema.gov/hiedu/docs/terms%20and%20definitions/Terms%20and%20Definitions.pdf
- DHS Risk Lexicon. http://www.dhs.gov/dhs-risk-lexicon
- European Security (Research) Glossary. http://www.focusproject.eu/web/focus/wiki/-/wiki/ESG/FrontPage
- FEMA Full Spectrum Risk Knowledgebase. https://www.fema.gov/office-national-capital-region-coordination-0/full-spectrum-risk-knowledgebase
- Institute for Crisis, Disaster, and Risk Management (ICDRM)/The George Washington University (GWU). "Emergency Management Glossary of Terms." Washington, D.C., June 30, 2010. http://www.gwu.edu/~icdrm/publications/PDF/GLOSSARY%20-%20Emergency%20Management%20ICDRM%2030%20JUNE%2010.pdf
- Urban Securipedia. http://www.securipedia.eu
- U.S. Department of Homeland Security—Definition of Terms. http://www.dhs.gov/definition-terms

Bibliographies

- Center for Strategic and International Studies (CSIS). "Counterterrorism Research Bibliography." Washington, D.C., n.d. Last accessed March 24, 2015. http://csis.org/images/stories/hrs/counterterrorismbibliography.pdf
- Federation of American Scientists. "Congressional Research Service Reports on Homeland Security." [Updated on a regular basis.] Last accessed March 24, 2015. http://www.fas.org/sgp/crs/homesec
- PRIO *New Security Studies* Book Series [listing of titles with content descriptions]. Last accessed March 24, 2015. http://www.routledge.com/books/series/PRIONEWSEC
- U.S. Army War College Library. "Homeland Security: A Selected Bibliography." Carlisle, PA, 2011. http://www.carlisle.army.mil/library/bibs/homesec11.pdf

Digital Libraries and Databases

- Centers for Disease Control and Prevention: Data & Statistics. http://www.cdc.gov/DataStatistics
- CIA World Factbook. https://www.cia.gov/library/publications/the-world-factbook
- EM-DAT. The International Disaster Database (Centre for Research on the Epidemiology of Disasters, CRED). http://www.emdat.be
- FEMA Emergency Management Institute—Educational Resources and Institutional Resources. http://training.fema.gov/hiedu
- Global Terrorism Database. http://www.start.umd.edu/gtd
- Homeland and Civil Security Collection (Penn State ScholarSphere). https://scholarsphere.psu.edu/collections/6395wb14x
- Homeland Security Digital Library (Naval Postgraduate School, Center for Homeland Defense and Security). http://www.hdsl.org
- Homeland Security Standards Database (American National Standards Institute, ANSI). http://www.hssd.us
- International Relations and Security Network (ISN) (Swiss Federal Institute of Technology Zurich). http://www.isn.ethz.ch
- International Security and Counter-Terrorism Reference Center. http://www.ebscohost.com/government/international-security-counter-terrorism-reference-center

- The National Academies Press, Conflict and Security Issues [prepublications, many of which address homeland and civil security research-related topics—such as resilience, security technologies, or ethics, legal, and social issues—available for free download]. http://www.nap.edu/topic/280/conflict-and-security-issues
- U.S. Fire Administration Database. http://www.usfa.fema.gov/data
- World Disasters Report. http://www.ifrc.org/publications-and-reports/world-disasters-report

Open Online Courses

- FEMA Emergency Management Institute—Independent Study Courses. https://training.fema.gov/is
- Naval Postgraduate School, Center for Homeland Defense and Security—Homeland Security Self-Study Program. https://www.chds.us/olc

Research Projects and Consortia Overviews

- European Union security research projects. http://cordis.europa.eu/fp7/security/projects_en.html
- FEMA Strategic Foresight Initiative. https://www.fema.gov/strategic-foresight-fema
- U.S. Department of Homeland Security, Centers of Excellence. http://www.dhs.gov/science-and-technology/centers-excellence
- U.S. Department of Homeland Security, Science and Technology Directorate—Project. https://www.dhs.gov/science-and-technology/fact-sheets

Trends, Policy Analyses, and News Sources

- Domestic Preparedness. http://www.domesticpreparedness.com
- Homeland Security News Wire. http://www.homelandsecuritynewswire.com
- HSTODAY.US. Insight and Analysis for Government Decision Makers. http://www.hstoday.us
- iHLS—Israel's Homeland Security Home. http://i-hls.com

Policy Core Documents

- U.S. and international core policy documents and further sources related to homeland and civil security are listed in the Penn State University Libraries' Homeland Security Research Guide that is updated on a regular basis: https://www.libraries.psu.edu/psul/researchguides/socialsciences/homeland_security.html

Handbooks

- Burgess, Peter, ed. *The Routledge Handbook of New Security Studies.* London: Routledge, 2010.

Offers a comprehensive theoretical and empirical overview of the evolving field of civil security in a critical perspective, in four key areas: new security concepts; new security subjects; new security objects; and new security practices. The focus is on the challenge of maintaining security in a globalized setting.

- Gill, Martin, ed. *The Handbook of Security.* 2nd ed. Basingstoke: Palgrave Macmillan, 2014.

Offers a comprehensive analysis of scholarly security debates as well as of current issues, informed by a range of disciplines, to include criminology, politics and economics, etc. Areas addressed comprise food security, cybersecurity, terrorism, organized crime, intelligence, surveillance, piracy, as well as analytical methods.

- Herrmann, Jeffrey W., ed. *Handbook of Operations Research for Homeland Security.* New York: Springer, 2013.

Addresses the state of the art in the application of operations research models related to terrorist attacks, natural disasters, and public health emergencies, and public and private sector responses.

- Kamien, David G., ed. *The McGraw-Hill Homeland Security Handbook: Strategic Guidance for a Coordinated Approach to Effective Security and Emergency Management.* 2nd ed. New York: McGraw-Hill, 2011.

Provides those in charge of homeland security and emergency preparedness in the public and in the private sector with a broad overview of critical homeland security issues, among other things including terrorism motivated by radical Islamist extremism; transportation and infrastructure protection; community and business resilience; intelligence and information; as well as international homeland security.

- Kemp, Roger L. *Homeland Security for the Private Sector: A Handbook.* Jefferson, NC: McFarland & Co., 2007.

Addresses new practices developed by private sector companies in order to protect their operations, employees, and customers in the current environment of threats to homeland security; collects best security practices from the private sector and includes a regional resource and a national resource directory.

- Voeller, John G., ed. *Wiley Handbook of Science and Technology for Homeland Security.* Hoboken, NJ: Wiley, 2010.

A research-based collection of resources on a broad range of areas of homeland security, reaching across disciplines and detailing the international dimension. Coverage includes cross-cutting themes and technologies (e.g., risk modeling and vulnerability assessment; sensing and detection; cybersecurity; human social and behavioral research; system and sector interdependencies; etc.) and key application areas.

Books

Foundations of Homeland Security

- Alperen, Martin J. *Foundations of Homeland Security: Law and Policy.* Hoboken, NJ: Wiley, 2011.

A guide to understanding the homeland security by its structure as shaped by underlying law and policy, covering statutes, policy papers, presidential directives, and other documents in a topic-by-topic format covering 25 different topic areas.

- Bullock, Jane A., George D. Haddow, and Damon P. Coppola. *Introduction to Homeland Security: Principles of All-Hazards Risk Management.* 4th ed. New York: Elsevier, 2013.

A well-established introduction to homeland security from the hazard and risk management point of view, including addressing of risk-informed priorities, including the cyber dimension, as well civil security research topics such as the safety-security continuum, communication, science and technology, and the future of homeland security.

- Ellis, John W. *Fundamentals of Homeland Security: An Operations Perspective.* Springfield, IL: Charles C. Thomas, 2014.

Covers homeland security from the implementation point of view, addressing mission, authority, organization, role, function, and fundamental terminology, focused on the need for organization to operate quickly and effectively, but under significant limitations, in emergencies. Includes a chapter on risk analysis with checklists summarizing relevant U.S. national plans and guidelines for threat, vulnerability, and consequence assessment.

- Ferro, Carme, David Henry, and Thomas MacLellan. *A Governor's Guide to Homeland Security.* Washington, D.C.: National Governors Association (NGA), NGA Center for Best Practices, Homeland Security & Public Safety Division, November 2010. http://www.nga.org/files/live/sites/NGA/files/pdf/1011GOVGUIDEHS.PDF

Introduces to the state-level implementation of homeland security policies and programs, and to state-level responsibilities for the security of the citizens, structured along capabilities to prepare, prevent, respond, and recover, with several sector specific examples for each.

- Marion, Nancy E., Kelley A. Cronin, and Willard M. Oliver. *Homeland Security: Policy and Politics.* Durham, NC: Carolina Academic Press, 2015.

A comprehensive analysis of controversial issues in the new and still emerging field of homeland security, focused on law enforcement and emergency management aspects. Topics covered include the history and future of FEMA; DHS funding; fusion centers; intelligence-led policing; evolution of the terrorist threat; role of citizen volunteers in disaster response; cyber-security; immigration policy; transportation network safety; etc.

- Martin, Gus. *Understanding Homeland Security.* Los Angeles, CA: Sage, 2015.

A comprehensive book that addresses subject areas such as emergency management, terrorism, criminal justice administration, intelligence, armed conflict, and social environments, with stimulation of critical thinking. Reviews theories, agency missions, laws, and regulations governing the U.S. homeland security enterprise.

- Morag, Nadav. *Comparative Homeland Security: Global Lessons.* Hoboken, NJ: John Wiley & Sons, 2011.

Provides a look beyond U.S. borders in terms of searching for solutions to existing homeland security policy problems, examining other countries' practices and allowing readers to integrate counterterrorism, emergency response and other practices from around the world into own policies. Coverage includes combating terrorism, countering radicalization, emergency response, border and transportation security, critical infrastructure protection, public health and military support of civil authorities.

- Oliver, Willard M., Nancy E. Marion, and Joshua B. Hill. *Introduction to Homeland Security: Policy, Organization, and Administration.* Burlington, MA: Jones & Bartlett, 2015.

Geared to the entry level and supported by an extensive online resource space, provides an accessible overview of homeland security definitions, policies, administrations, and organizations, addressing foundational issues as well as policy concepts and political, legal, and technological responses to the contemporary challenges.

- Ramsay, James D., and Linda Kiltz, eds. *Critical Issues on Homeland Security: A Casebook.* Boulder, CO: Westview, 2014.

A collection of homeland and civil security research case studies, the book encourages analytical examination and critical thinking related to complex practical homeland security problems, covering a broad range of legal, policy, and operational challenges and cases on such issues as arctic security, use of drones, cybersecurity, and lessons of the Mumbai attacks of 2008 and the Haiti earthquake of 2010.

- Sylves, Richard. *Disaster Policy and Politics: Emergency Management and Homeland Security.* 2nd ed. Washington, D.C.: CQ Press, 2015.

Covers emergency management as a part of the homeland security mission space, using an all-hazards approach and public management as well as historical, legal, and science and engineering perspectives, added to by mini-case studies from the United States and some other countries.

Foundations of Civil Security Research

- Fischer, Robert J., Edward P. Halibozek, and David C. Walters. *Introduction to Security.* 9th ed. Waltham, MA: Butterworth-Heinemann, 2013.

Provides a comprehensive overview of civil security, with a focus on private security, but also including tracking of the origins and development of security in the 21st century. A chapter is dedicated to the role of private enterprises in homeland security, as well as to relevant legal foundations, including private security powers, liability, and privacy. Addresses risk analyses, perimeter protection, contingency planning, and specific threats and solutions.

- Siedschlag, Alexander, ed. *Scenario-Based Security Foresight.* Special Issue, *Information & Security* 29, nos. 1–2 (2013). http://procon.bg/29-scenario-based-security-foresight

Discusses methods and techniques in scenario-based foresight; presents selected empirical results from scenario foresight on threats, risk management needs, and future European Union roles as a comprehensive security provider; links the scenarios to requirements for security research and industry products.

- Smith, Clifton L., and David J. Brooks: *Security Science: The Theory and Practice of Security.* New York: Elsevier, 2013.

Integrates multidisciplinary practice areas of security into a single body of knowledge, based on an evidence-based approach as well as consideration of theories, principles, and models or frameworks. The scope reaches from organizational security to homeland security. The focus is on security and risk management, knowledge management, business continuity management, and technological solutions for security problems.

- Thoma, Klaus, ed. *European Perspectives on Security Research.* Munich: acatech—Deutsche Akademie der Technikwissenschaften, 2011. http://www.acatech.de/fileadmin/user_upload/Baumstruktur_nach_Website/Acatech/root/de/Publikationen/Englisch/acatech_diskutiert_European-Research_WEB.pdf

Represents the science and engineering perspective on civil security research in Europe, including the Russian Federation, with a focus on the safety-security continuum, risk and vulnerability assessment, and critical infrastructure protection, also addressing security vs. liberty considerations.

- Zedner, Lucia. *Security.* London: Routledge, 2009.

Provides a conceptual discussion of the creation of security as a public good from a criminological perspective but in an interdisciplinary fashion, addressing the erosion of distinctions between public safety and private protection, policing and security services, as well as national and international security and post-9/11 securitization trends.

Essential Further Reading

- Balzacq, Thierry, ed. *Securitization Theory: How Security Problems Emerge and Dissolve.* London: Routledge, 2010.

Provides a new conceptual framework for the analysis of securitization processes, increasing understanding of the emergence, evolution, and dissolution of security issues. Examples used include terrorism, pandemics, and environmental issues. Methods used include discourse analysis, process-tracing, ethnographic research, and content analysis.

- Busch, Nathan E., and Austen D. Givens. *The Business of Counterterrorism: Public-Private Partnerships in Homeland Security.* New York: Peter Lang, 2014.

An account of the opportunities and challenges that public-private partnerships face in the post-9/11 world, focusing on critical infrastructure protection; cybersecurity; information sharing; security at U.S. ports of entry; and disaster recovery.

- Egli, Dane S. *Beyond the Storms: Strengthening Homeland Security and Disaster Management to Achieve Resilience.* Armonk, NY and London: Sharpe, 2014.

Deals with all-hazard disaster preparedness of the United States and its people, proposing a comprehensive approach with a focus on critical infrastructure investments and risk assessment, and including a holistic analysis and implementation of mitigation, response, and recovery efforts.

- Flammini, Francesco, Roberto Setola, and Giorgio Franceschetti, eds. *Effective Surveillance for Homeland Security: Balancing Technology and Social Issues.* Boca Raton, FL: CRC Press, 2013.

Addresses, in an international perspective, the challenge of balancing technology capabilities and social issues in surveillance applications for homeland security, with a multidisciplinary focus on the societal dimension of surveillance; advanced technologies for surveillance; and applications of surveillance systems in the framework of homeland security. The book includes real-world case studies but also examines cutting-edge research topics.

- *Issues in Terrorism and Homeland Security: Selections from CQ Researcher.* Thousand Oaks, CA: Sage, 2011.

Offers a comprehensive perspective including critical thinking on understanding and countering terrorism, including topics such as: Could 9/11 have been avoided?; anti-Americanism and hate groups; terrorism and the Internet; understanding Islam; global Jihad; homeland security and human rights; policing and prosecution; etc.

- Kapucu, Naim, Christopher V. Hawkins, and Fernando I. Rivera. *Disaster Resiliency: Interdisciplinary Perspectives.* New York: Routledge, 2013.

Focused on natural disasters, offers an international array of systematic and empirical assessments of resiliency, employing a variety of statistical techniques and comparative methods that address such aspects as public management and policy, community planning and development, and cultural and social capabilities for resiliency.

- Kemp, Roger L. ed., *Homeland Security: Best Practices for Local Government.* Washington, D.C.: International City/County Management Association, 2010.

Structured along the homeland security management cycle (mitigation, preparedness, response, and recovery), offers a comprehensive resource on homeland security best practices for local government, including case studies and citizen-involving strategies. Intelligence, information management, communication, public health, vulnerability assessment, regional resilience, stress management, and other aspects are covered.

- Kettl, Donald F. *System under Stress: Homeland Security and American Politics.* London: Sage, 2014.

Looks at U.S. homeland security as a complex adaptive system that needs to adopt a risk management approach, accepting that no "total protection" is possible, but that therefore there is "zero balance for mistakes" in homeland security political programs and practical strategies. The book discusses how to achieve this in an all-hazards context and in the face of coordination dilemmas and the challenge of bureaucracy.

- McEntire, David E., ed. *Comparative Emergency Management: Understanding Disaster Policies, Organizations, and Initiatives from Around the World.* E-book, FEMA Emergency Management Institute, 2013. http://training.fema.gov/hiedu/aemrc/booksdownload/compemmgmtbookproject

Aiming to increase knowledge of international emergency management systems, the book identifies the importance of comparison and then addresses approaches and challenges in countries across the world, establishing a background for improved lessons learned processes and outcomes.

- Perrow, Charles. *The Next Catastrophe: Reducing Our Vulnerabilities to Natural, Industrial, and Terrorist Disasters.* Princeton, NJ: Princeton University Press, 2007.

Expands on that author's ideas about normal accidents, that is the notion that multiple and unexpected failures are built into our society's complex systems. Suggests a shift in focus from protection to minimization of damage and attractiveness of targets to terrorists.

- Steiner, James E. *Homeland Security Intelligence.* Los Angeles, CA: Sage, 2015.

A comprehensive treatment of intelligence with the full range of homeland security practitioners at state and local levels and in the private sector in mind, with a focus on counterterrorism and cybersecurity.

- Vermeulen, Gert, and Wendy De Bondt. *Justice, Home Affairs and Security: European and International Institutional and Policy Development.* Antwerpen: Maklu, 2015.

Covers the historical, institutional, and topical development of European Union policy in the areas of justice, home affairs, and security, embedded in a broader international context including the Council of Europe, NATO, OSCE, G8/G7, OECD, and the United Nations.

- Wuthnow, Robert. *Be Very Afraid: The Cultural Response to Terror, Pandemics, Environmental Devastation, Nuclear Annihilation, and Other Threats.* Oxford: Oxford University Press, 2010.

Applies the cultural approach to explain the institutionalization of existential fear in a large part of United Sates' public life and public policy over time, before and in the era of homeland security—along with a drive for action at almost any price that is likely to miss its purpose of making citizens need to be less afraid.

Journals

Homeland Security

- *Counter Terrorist Magazine.* http://www.thecounterterroristmag.com

An American informational journal for law enforcement, intelligence, and special operations professionals, focused on dissemination of operational experience and topics such as terrorist tactics, techniques, and procedures; organized crime; policing and security operations; border security; emergency and medical services; technology and science; and cultural study.

- *Domestic Preparedness Journal.* http://www.domesticpreparedness.com/DomPrep_Journal

A monthly online professional journal the for multidiscipline and multi-jurisdictional preparedness community, including first responders, medical receivers, emergency planners, local-state-federal authorities, non-government organizations and private-sector professionals, including a focus on state-level homeland security.

- *Homeland Defense and Civil Support Journal.* http://www.csl.army.mil/homelanddefensejournal/JournalHome.aspx

Strategic, futuristic, and innovative scholarly and practical contributions to homeland defense, homeland security, and defense support of civil authorities.

- *Homeland Security Affairs.* http://www.hsaj.org

Journal of the Naval Postgraduate School Center for Homeland Defense and Security (CHDS), with analysis and discussion of strategies, policies, and organizational contexts of U.S. homeland security.

- *Homeland Security Today.* http://www.hstoday.us

A monthly print magazine targeting officials with homeland security responsibilities at all levels of government, as well as executives in the private sector, with themes centered on practical and product aspects related to protecting from terrorism and preparing for all hazards.

- *Inside Homeland Security.* http://www.abchs.com/ihs

A monthly online magazine by the American Board for Certification in Homeland Security, with articles that provide in-depth information from homeland security authorities, as well as articles on current events, continuing education, and certifications. Main topics include counterterrorism, intelligence, weapons of mass destruction, critical incident management, law enforcement, and first responders.

- *Journal of Homeland Security Education.* http://www.journalhse.org

An academic journal focused on educational practice of homeland security, addressing issues of homeland security pedagogy from the associate to doctoral-degree level, facilitating exchange of ideas for teaching and learning in an emerging academic discipline of homeland security.

- *Journal of Homeland Security and Emergency Management.* http://www.degruyter.com/view/j/jhsem

An interdisciplinary academic journal promoting a comprehensive and dynamic perspective on homeland security and on emergency management as it is part of the homeland security mission space, serving as a bridge between the two fields, and between communities of researchers and practitioners to foster best practices.

Civil Security Research

- *Safety & Security International.* http://www.monch.com/mpg/publications/safety-a-security-international

A bimonthly magazine providing first-hand reports, analyses, evaluations, and updates on developments and trends in the global security environment, focusing on the interaction of new technologies with policies, strategies, and programs. The focus is on border control; critical infrastructures; maritime security; nuclear, biological, and chemical detection and protection; airport, port, and transport security; and disaster management.

- *Security Dialogue.* http://sdi.sagepub.com

Aims to combine advances in theory with results from empirical research from the whole bandwidth of the field of studies of security, including civil security, with perspectives from across different disciplines such as gender studies, political sociology, political economy, political theory, international relations, anthropology, psychoanalysis, political philosophy, etc.

- *Security Journal.* http://www.palgrave-journals.com/sj

Provides information about latest developments and techniques in security management across fields and sectors, from crime to critical infrastructure, addressing security researchers and professionals alike.

Related Disciplines

- *Behavioral Sciences of Terrorism and Political Aggression.* http://www.tandfonline.com/rirt
- *Critical Studies on Terrorism.* http://www.tandfonline.com/rter
- *Democracy and Security.* http://tandfonline.com/fdas
- *Disaster Medicine and Public Health Preparedness.* http://journals.cambridge.org/action/displayJournal?jid=DMP
- *Disaster Prevention and Management.* http://www.emeraldinsight.com/journal/dpm
- *Disasters. The Journal of Disaster Studies, Policy and Management.* http://onlinelibrary.wiley.com/journal/10.1111/%28ISSN%291467-7717
- *Harvard National Security Journal.* http://harvardnsj.org
- *Intelligence and National Security.* http://www.tandfonline.com/fint
- *International Journal of Business Continuity and Risk Management.* http://www.inderscience.com/jhome.php?jcode=ijbcrm

- *International Journal of Disaster Risk Science.* http://link.springer.com/journal/13753
- *International Journal of Emergency Management.* http://www.inderscience.com/jhome.php?jcode=ijem
- *International Journal of Intelligence and Counterintelligence.* http://www.tandfonline.com/ujic
- *International Journal of Mass Emergencies and Disasters.* http://www.ijmed.org
- *Journal of Business Continuity & Emergency Planning.* http://www.henrystewartpublications.com/jbcep
- *Journal of Contingencies and Crisis Management.* http://onlinelibrary.wiley.com/journal/10.1111/%28ISSN%291468-5973
- *Journal of Emergency Management.* http://ejournal.emergencymanagementjournal.com
- *Journal of National Security Law and Policy.* http://jnslp.com
- *Journal of Policing, Intelligence and Counter Terrorism.* http://www.tandfonline.com/rpic
- *National Security Law Journal.* http://jnslp.com
- *Resilience: International Policies, Practices and Discourses.* http://www.tandfonline.com/resi
- *Risk Analysis.* http://onlinelibrary.wiley.com/journal/10.1111/%28ISSN%291539-6924
- *Studies in Conflict and Terrorism.* http://www.tandfonline.com/uter
- *Studies in Intelligence.* https://www.cia.gov/library/center-for-the-study-of-intelligence/csi-publications/csi-studies
- *Terrorism and Political Violence.* http://www.tandfonline.com/ftpv

List of Contributors

Todd S. Bacastow, Ph.D., is Professor of Practice for Geospatial Intelligence in The Pennsylvania State University's College of Earth and Mineral Sciences and Director of the Geospatial Intelligence Program.

Michel F. Bosco, Ph.D., leads, since October 2013, the European Commission program for Security Research, and is the Deputy Head of the Unit in charge of Innovation and Industry for Security.

Rodger Broomé, Ph.D., is a retired police officer and firefighter/EMT from West Jordan, Utah, USA and holds a faculty position at Utah Valley University.

Dane S. Egli, Ph.D., is a National Security and Homeland Security senior advisor at Johns Hopkins University Applied Physics Laboratory in Laurel, Maryland.

John R. Fisher, Ph.D., is an Associate Professor in the Emergency Management Bachelor's program and Director of the Institute of Emergency Services and Homeland Security at Utah Valley University.

John Guelke, Ph.D., is a research fellow at the University of Warwick's (UK) Interdisciplinary Ethics Research Group.

Katerina Hadjimatheou, Ph.D., is a research fellow at the University of Warwick's (UK) Interdisciplinary Ethics Research Group.

Andrea Jerković, Ph.D., is Director of CEUSS | Center for European Security Studies at Sigmund Freud University Vienna, Austria, and an executive member of the Kölner Forum für Internationale Beziehungen und Sicherheitspolitik e.V. (KFIBS)/Cologne Forum for International Relations and Security Policy.

R. Jeffery Maxfield, Ed.D., is a Professor in the Emergency Services Department and recently served as Associate Dean over the School of Public Services and Director of the Institute of Emergency Services and Homeland Security at Utah Valley University.

Robert McCreight, Ph.D., is a Senior Consultant with Global Concepts & Communications, LLC. He has over 35 years' experience in the State Department working on global security, arms control, intelligence operations, bio warfare, nuclear weaponry, counterterrorism, emergency humanitarian missions, and POLMIL affairs.

Lloyd Mitchell, Ph.D., is an Associate Professor at Elizabeth City State University. He also has collaborated with the Department of Homeland Security (DHS) on several projects related to challenges and issues involving American Indians and tribal entities.

Kevin J. Molloy, MEP, retired from the U.S. Department of Homeland Security Federal Emergency Management Agency (FEMA) National Integration Center as Senior Project Officer. Molloy has over 40 years' experience in the Emergency Management field at the local, state, and federal levels. He served as Dauphin County coordinator during the Three Mile Island Nuclear Accident in 1979.

Nadav Morag, Ph.D., teaches at the Center for Homeland Defense and Security (CHDS), United States Naval Postgraduate School in Monterey, California. He concurrently serves as University Dean of the School of Security Studies at Colorado Tech and is a Senior Fellow at the Center for Cyber and Homeland Security at The George Washington University.

Steven A. Peterson, Ph.D., was Director of the School of Public Affairs at The Pennsylvania State University, Penn State Harrisburg, from 1997 to 2015, where he is Professor of Politics and Public Affairs.

William J. Ryan, MBA, is a U.S. Department of Homeland Security as a Protective Security Advisor (PSA). Prior to assuming his PSA duties, he served as an intelligence officer in the CIA's National Clandestine Service (formerly the Directorate of Operations).

Alexander Siedschlag, Ph.D., is Professor of Homeland Security and Public Health Preparedness at The Pennsylvania State University, Penn State Harrisburg, in a joint appointment with the Department of Public Health Sciences at the Hershey School of Medicine. He is Chair of Penn State Homeland Security Programs, including the Intercollege Master of Professional Studies in Homeland Security (*i*MPS-HLS).

Tom Sorell, Ph.D., is Professor of Politics and Philosophy and Head of the Interdisciplinary Ethics Research Group in the Department of Politics and International Studies (PAIS) at the University of Warwick, UK.

Rachel Suissa, Ph.D., is an Adjunct Professor at the National Security Program at the School of Political Sciences at the University of Haifa, Israel, and a Senior Research Fellow at the Haifa Center for German and European Studies.

Gregory A. Thomas, Ph.D., is Professor of Practice and the Assistant Director of Geospatial Intelligence Programs in the John A. Dutton e-Education Institute in the College of Earth and Mineral Sciences with the Pennsylvania State University.

Index

D

E

F

G

H

I

J

K

L

M

N

O

P

Q

R

S

T

U

V

W